NUREG-1806, Vol. 2

Technical Basis for Revision of the Pressurized Thermal Shock (PTS) Screening Limit in the PTS Rule (10 CFR 50.61)

Appendices

Manuscript Completed: May 2006
Date Published: August 2007

Prepared by
M. EricksonKirk, M. Junge, W. Arcieri,
B.R. Bass, R. Beaton, D. Bessette,
T.H.J. Chang, T. Dickson, C.D. Fletcher,
A. Kolaczkowski, S. Malik, T. Mintz,
C. Pugh, F. Simonen, N. Siu,
D. Whitehead, P. Williams, R. Woods,
S. Yin

Division of Fuel, Engineering and Radiological Research
Office of Nuclear Regulatory Research
U.S. Nuclear Regulatory Commission
Washington, DC 20555-0001

Abstract

During plant operation, the walls of reactor pressure vessels (RPVs) are exposed to neutron radiation, resulting in localized embrittlement of the vessel steel and weld materials in the core area. If an embrittled RPV had a flaw of critical size and certain severe system transients were to occur, the flaw could very rapidly propagate through the vessel, resulting in a through-wall crack and challenging the integrity of the RPV. The severe transients of concern, known as pressurized thermal shock (PTS), are characterized by a rapid cooling of the internal RPV surface in combination with repressurization of the RPV. Advancements in our understanding and knowledge of materials behavior, our ability to realistically model plant systems and operational characteristics, and our ability to better evaluate PTS transients to estimate loads on vessel walls led the NRC to realize that the earlier analysis, conducted in the course of developing the PTS Rule in the 1980s, contained significant conservatisms.

This report summarizes 21 supporting documents that describe the procedures used and results obtained in the probabilistic risk assessment, thermal hydraulic, and probabilistic fracture mechanics studies conducted in support of this investigation. Recommendations on toughness-based screening criteria for PTS are provided.

Foreword

The reactor pressure vessel is exposed to neutron radiation during normal operation. Over time, the vessel steel becomes progressively more brittle in the region adjacent to the core. If a vessel had a preexisting flaw of critical size and certain severe system transients occurred, this flaw could propagate rapidly through the vessel, resulting in a through-wall crack. The severe transients of concern, known as pressurized thermal shock (PTS), are characterized by rapid cooling (i.e., thermal shock) of the internal reactor pressure vessel surface that may be combined with repressurization. The simultaneous occurrence of critical-size flaws, embrittled vessel, and a severe PTS transient is a very low probability event. The current study shows that U.S. pressurized-water reactors do not approach the levels of embrittlement to make them susceptible to PTS failure, even during extended operation well beyond the original 40-year design life.

Advancements in our understanding and knowledge of materials behavior, our ability to realistically model plant systems and operational characteristics, and our ability to better evaluate PTS transients to estimate loads on vessel walls have shown that earlier analyses, performed some 20 years ago as part of the development of the PTS rule, were overly conservative, based on the tools available at the time. Consistent with the NRC's Strategic Plan to use best-estimate analyses combined with uncertainty assessments to resolve safety-related issues, the NRC's Office of Nuclear Regulatory Research undertook a project in 1999 to develop a technical basis to support a risk-informed revision of the existing PTS Rule, set forth in Title 10, Section 50.61, of the Code of Federal Regulations (10 CFR 50.61).

Two central features of the current research approach were a focus on the use of realistic input values and models and an explicit treatment of uncertainties (using currently available uncertainty analysis tools and techniques). This approach improved significantly upon that employed in the past to establish the existing 10 CFR 50.61 embrittlement limits. The previous approach included unquantified conservatisms in many aspects of the analysis, and uncertainties were treated implicitly by incorporating them into the models.

This report summarizes a series of 21 reports that provide the technical basis that the staff will consider in a potential revision of 10 CFR 50.61; it includes a description of analysis procedures and a detailed discussion of findings. The risk from PTS was determined from the integrated results of the Fifth Version of the Reactor Excursion Leak Analysis Program (RELAP5) thermal-hydraulic analyses, fracture mechanics analyses, and probabilistic risk assessment. These calculations demonstrate that, even through the period of license extension, the likelihood of vessel failure attributable to PTS is extremely low ($\approx 10^{-8}$/year) for all domestic pressurized water reactors. Limited analyses are continuing to further evaluate this finding. Should the $\approx 10^{-8}$/year value be confirmed, this would provide a basis for significant relaxation, or perhaps elimination, of the embrittlement limit established in 10 CFR 50.61. Such changes would reduce unnecessary conservatism without affecting safety because the operating reactor fleet has little probability of exceeding the limits on the frequency of reactor vessel failure established from NRC guidelines on core damage frequency and large early release frequency through the period of license extension.

Brian W. Sheron, Director
Office of Nuclear Regulatory Research
U.S. Nuclear Regulatory Commission

Contents

Appendices

Figures

Tables

Executive Summary

This report summarizes the results of a 5-year study conducted by the U.S. Nuclear Regulatory Commission (NRC), Office of Nuclear Regulatory Research (RES). The aim of this study was to develop the technical basis for revision of the Pressurized Thermal Shock (PTS) Rule, as set forth in Title 10, Section 50.61, of the *Code of Federal Regulations* (10 CFR 50.61), "Fracture Toughness Requirements for Protection Against Pressurized Thermal Shock Events," consistent with the NRC's current guidelines on risk-informed regulation. This report, together with other supporting reports documenting the study details and results, provides this basis.

This executive summary begins with a description of PTS, how it might occur, and its potential consequences for the reactor pressure vessel (RPV). This is followed by a summary of the current regulatory approach to PTS, which leads directly to a discussion of the motivations for conducting this project. Following this introductory information, we describe the approach used to conduct the study, and summarize our key findings and recommendations, which include a proposal for revision of the PTS screening limits. We then conclude the executive summary with a discussion of the potential impact of this proposal on regulations other than 10 CFR 50.61.

Description of PTS

During the operation of a nuclear power plant, the RPV walls are exposed to neutron radiation, resulting in localized embrittlement of the vessel steel and weld materials in the area of the reactor core. If an embrittled RPV had an existing flaw of critical size and certain severe system transients were to occur, the flaw could propagate very rapidly through the vessel, resulting in a through-wall crack and challenging the integrity of the RPV. The severe transients of concern, known as pressurized thermal shock (PTS), are characterized by a rapid cooling (i.e., thermal shock) of the internal RPV surface and downcomer, which may be followed by repressurization of the RPV. Thus, a PTS event poses a potentially significant challenge to the structural integrity of the RPV in a pressurized-water reactor (PWR).

A number of abnormal events and postulated accidents have the potential to thermally shock the vessel (either with or without significant internal pressure). These events include a pipe break or stuck-open valve in the primary pressure circuit, a break of the main steam line, etc. During such events, the water level in the core drops as a result of the contraction produced by rapid depressurization. In events involving a break in the primary pressure circuit, an additional drop in water level occurs as a result of leakage from the break. Automatic systems and operators must provide makeup water in the primary system to prevent overheating of the fuel in the core. However, the makeup water is much colder than that held in the primary system. As a result, the temperature drop produced by rapid depressurization coupled with the near-ambient temperature of the makeup water produces significant thermal stresses in the thick section steel wall of the RPV. For embrittled RPVs, these stresses could be sufficient to initiate a running crack, which could propagate all the way through the vessel wall. Such through-wall cracking of the RPV could precipitate core damage or, in rare cases, a large early release of radioactive material to the environment. Fortunately, the coincident occurrence of critical-size flaws, embrittled vessel steel and weld material, and a severe PTS transient is a very low-probability event. In fact, only a few currently operating PWRs are projected to closely approach the current statutory limit on the level of embrittlement during their planned operational life.

Current Regulatory Approach to PTS

As set forth in 10 CFR 50.61, the PTS Rule requires licensees to monitor the embrittlement of their RPVs using a reactor vessel material surveillance program qualified under Appendix H to 10 CFR Part 50, "Reactor Vessel Material Survellience Program Requirements." The surveillance results are then used together with the formulae and tables in 10 CFR 50.61 to estimate the fracture toughness transition temperature (RT_{NDT}) of the steels in the vessel's beltline and how those transition temperatures increase as a result of irradiation damage throughout the operational life of the vessel. For licensing purposes, 10 CFR 50.61 provides instructions on how to use these estimates of the effect of irradiation damage to estimate the value of RT_{NDT} that will occur at end of license (EOL), a value called RT_{PTS}. 10 CFR 50.61 also provides "screening limits" (maximum values of RT_{NDT} permitted during the plant's operational life) of +270°F (132°C) for axial welds, plates, and forgings, and +300°F (149°C) for circumferential welds. These screening limits correspond to a limit of $5x10^{-6}$ events/year on the annual probability of developing a through-wall crack [RG 1.154]. Should RT_{PTS} exceed these screening limits, 10 CFR 50.61 requires the licensee to either take actions to keep RT_{PTS} below the screening limit (by implementing "reasonably practicable" flux reductions to reduce the embrittlement rate, or by deembrittling the vessel by annealing [RG 1.162]), or perform plant-specific analyses to demonstrate that operating the plant beyond the 10 CFR 50.61 screening limit does not pose an undue risk to the public [RG 1.154].

While no currently operating PWR has an RT_{PTS} value that exceeds the 10 CFR 50.61screening limit before EOL, several plants are close to the limit (3 are within 2°F, while 10 are within 20°F). Those plants are likely to exceed the screening limit during the 20-year license renewal period that is currently being sought by many operators. Moreover, some plants maintain their RT_{PTS} values below the 10 CFR 50.61 screening limits by implementing flux reductions (low-leakage cores, ultra-low-leakage cores), which are fuel management strategies that can be economically deleterious in a deregulated marketplace. Thus, the 10 CFR 50.61 screening limits can restrict both the licensable and economic lifetime of PWRs.

Motivation for this Project

It is now widely recognized that the state of knowledge and data limitations in the early 1980s necessitated conservative treatment of several key parameters and models used in the probabilistic calculations that provided the technical basis for the current PTS Rule. The most prominent of these conservatisms include the following factors:

- highly simplified treatment of plant transients (very coarse grouping of many operational sequences (on the order of 10^5) into very few groups ($\approx$10), necessitated by limitations in the computational resources needed to perform multiple thermal-hydraulic calculations)
- lack of any significant credit for operator action
- characterization of fracture toughness using RT_{NDT}, which has an intentional conservative bias
- use of a flaw distribution that places *all* flaws on the interior surface of the RPV, and, in general, contains larger flaws than those usually detected in service
- a modeling approach that treated the RPV as if it were made entirely from the most brittle of its constituent materials (welds, plates, or forgings)
- a modeling approach that assessed RPV embrittlement using the peak fluence over the entire interior surface of the RPV

These factors indicate the high likelihood that the current 10 CFR 50.61 PTS screening limits are unnecessarily conservative. Consequently, the NRC staff believed that reexamining the technical basis for these screening limits, based on a modern understanding of all the factors that influence PTS, would most likely provide strong justification for substantially relaxing these limits. For these reasons, RES undertook this study with the objective of developing the technical basis to support a risk-informed revision of the PTS Rule and the associated PTS screening limit.

Approach

As illustrated in the following figure, three main models (shown as solid blue squares), taken together, allow us to estimate the annual frequency of through-wall cracking in an RPV:

- probabilistic risk assessment (PRA) event sequence analysis
- thermal-hydraulic (TH) analysis
- probabilistic fracture mechanics (PFM) analysis

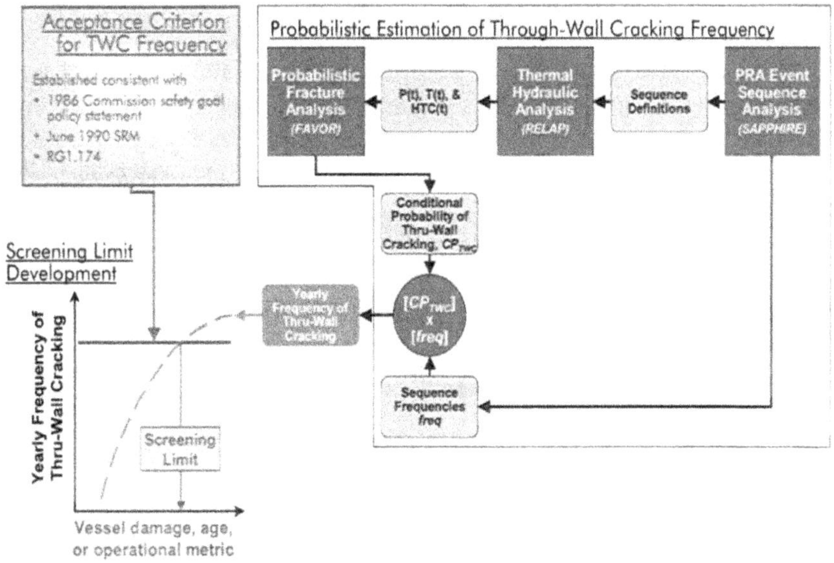

Schematic showing how a probabilistic estimate of through-wall cracking frequency (TWCF) is combined with a TWCF acceptance criterion to arrive at a proposed revision of the PTS screening limit

First, a PRA event sequence analysis is performed to define the sequences of events that are likely to cause a PTS challenge to RPV integrity, and estimate the frequency with which such sequences can be expected to occur. The event sequence definitions are then passed to a TH model that estimates the temporal variation of temperature, pressure, and heat-transfer coefficient in the RPV downcomer, which is characteristic of each sequence definition. These temperature, pressure, and heat-transfer coefficient histories are then passed to a PFM model that uses the TH output, along with other information concerning plant design and construction materials, to estimate the time-dependent "driving force to fracture" produced by a particular event sequence. The PFM model then compares this estimate of fracture driving force to the fracture toughness, or fracture resistance, of the RPV steel. This comparison allows us to estimate the probability that a crack could grow to sufficient size that it would penetrate all the way through the RPV wall if that particular sequence of events actually occured. The final step in the analysis involves a simple matrix multiplication of the probability of through-wall cracking (from the PFM analysis) with the frequency at which a particular event sequence is expected to occur (as defined by the event-tree analysis). This product establishes an estimate of the annual frequency of through-wall cracking that can be expected for a particular plant after a particular period of operation when subjected to a particular sequence of events. The annual frequency of through-wall cracking is then summed for all event sequences to estimate the total annual frequency of through-wall cracking for the vessel. Performance of such analyses for various operating lifetimes provides an estimate of how the annual frequency of through-wall cracking can be expected to vary over the lifetime of the plant.

The probabilistic calculations just described are performed to establish the technical basis for a revised PTS Rule within an integrated systems analysis framework. Our approach considers a broad range of factors that influence the likelihood of vessel failure during a PTS event, while accounting for uncertainties in these factors across a breadth of technical disciplines. Two central features of this approach are a focus on the use of realistic input values and models (wherever possible), and an *explicit* treatment of uncertainties (using currently available uncertainty analysis tools and techniques). Thus, our current approach improves upon that employed in developing SECY-82-465, which included intentional and unquantified conservatisms in many aspects of the analysis, and treated uncertainties *implicitly* by incorporating them into the models.

Key Findings

The findings from this study are divided into the following five topical areas: (1) the expected magnitude of the through-wall cracking frequency (TWCF) for currently anticipated operational lifetimes, (2) the material factors that dominate PTS risk, (3) the transient classes that dominate PTS risk, (4) the applicability of these findings (based on detailed analyses of three PWRs) to PWRs *in general*, and (5) the annual limit on TWCF established consistent with current guidelines on risk-informed regulation. In this summary, ***the conclusions are presented in boldface italic***, while the supporting information is shown in regular type.

TWCF Magnitude for Currently Anticipated Operational Lifetimes

- ***The degree of PTS challenge is low for currently anticipated lifetimes and operating conditions.***
  - Even at the end of license extension (60 operational years, or 48 effective full-power years (EFPY) at an 80% capacity factor), the mean estimated TWCF does not exceed $2x10^{-8}$/year for the plants analyzed. Considering that the RPVs at the Beaver Valley Power Station and Palisades Nuclear Power Plant are constructed from some of the most irradiation-sensitive materials in commercial reactor service today, these results suggest that, provided that operating practices do not change dramatically in the future, the operating reactor fleet is in little danger of exceeding either the TWCF limit of $5x10^{-6}$/yr expressed by Regulatory Guide 1.154 [RG 1.154] or the value of $1x10^{-6}$/yr recommended in Chapter 10 of this report — even after license extension.

Material Factors and their Contributions to PTS Risk

- ***Axial flaws, and the toughness properties that can be associated with such flaws, control nearly all of the TWCF.***
  - Axial flaws are much more likely than circumferential flaws to propagate through the RPV wall because the applied fracture driving force increases continuously with increasing crack depth for an axial flaw. Conversely, circumferentially oriented flaws experience a driving force peak mid-wall, providing a natural crack arrest mechanism. It should be noted that crack initiation from circumferentially oriented flaws is likely; it is only their through-wall propagation that is much less likely (relative to axially oriented flaws).
  - It is, therefore, the toughness properties that can be associated with axial flaws that control nearly all of the TWCF. These include the toughness properties of plates and axial welds at the flaw locations. Conversely, the toughness properties of both circumferential welds and forgings have little effect on the TWCF because these can be associated only with circumferentially oriented flaws.

Transients and their Contributions to PTS Risk

- ***Transients involving primary side faults are the dominant contributors to TWCF, while transients involving secondary side faults play a much smaller role.***
  - The severity of a transient is controlled by a combination of three factors:
    - initial cooling rate, which controls the thermal stress in the RPV wall
    - minimum temperature of the transient, which controls the resistance of the vessel to fracture
    - pressure retained in the primary system, which controls the pressure stress in the RPV wall
  - The significance of a transient (i.e., how much it contributes to PTS risk) depends on these three factors and the likelihood that the transient will occur.
  - Our analysis considered transients in the following classes (as shown in the following table):
    - primary side pipe breaks
    - stuck-open valves on the primary side
    - main steam line breaks
    - stuck-open valves on the secondary side
    - feed-and-bleed
    - steam generator tube rupture
    - mixed primary and secondary initiators

Factors contributing to the severity and risk-dominance of various transient classes

Transient Class		Transient Severity			Transient Likelihood	TWCF Contribution
		Cooling Rate	Minimum Temperature	Pressure		
Primary Side Pipe Breaks	Large-Diameter	Fast	Low	Low	Low	Large
	Medium-Diameter	Moderate	Low	Low	Moderate	Large
	Small-Diameter	Slow	High	Moderate	High	~0
Stuck-Open Valves, Primary Side	Valve Recloses	Slow	Moderate	High	High	Large
	Valve Remains Open	Slow	Moderate	Low	High	~0
Main Steam Line Break		Fast	Moderate	High	High	Small
Stuck-Open Valve(s), Secondary Side		Moderate	High	High	High	~0
Feed-and- Bleed		Slow	Low	Low	Low	~0
Steam Generator Tube Rupture		Slow	High	Moderate	Low	~0
Mixed Primary & Secondary Initiators		Slow	Mixed		Very Low	~0
Color Key		Enhances TWCF Contribution	Intermediate		Diminishes TWCF Contribution	

  - The table above provides a qualitative summary our results for these transient classes in terms of both transient severity and the likelihood that the transient will occur. The color-coding of table entries indicates the contribution (or lack thereof) of these factors to the TWCF of the various classes of transients. This summary indicates that the risk-dominant transients (medium- and large-diameter primary side pipe breaks, and stuck-open primary side valves that later reclose) all have multiple factors that, in combination, result in their significant contributions to TWCF.
    - For medium- to large-diameter primary side pipe breaks, the fast to moderate cooling rates and low downcomer temperatures (generated by rapid depressurization and emergency injection of low-temperature makeup water directly to the primary) combine to produce a high-severity transient. Despite the moderate to low likelihood that these transients will occur, their severity (if they do occur) makes them significant contributors to the total TWCF.

- For stuck-open primary side valves that later reclose, the repressurization associated with valve reclosure coupled with low temperatures in the primary combine to produce a high-severity transient. This, coupled with a high likelihood of transient occurrence, makes stuck-open primary side valves that later reclose significant contributors to the total TWCF.
- The small or negligible contribution of all secondary side transients (main steam line break, stuck-open secondary valves) results directly from the lack of low temperatures in the primary system. For these transients, the minimum temperature of the primary for times of relevance is controlled by the boiling point of water in the secondary (212°F (100°C) or above). At these temperatures, the fracture toughness of the RPV steel is sufficiently high to resist vessel failure in most cases.

Applicability of These Findings to PWRs in General

- ***Credits for operator action, while included in our analysis, do not influence these findings in any significant way.*** Operator action credits can dramatically influence the risk-significance of ***individual*** transients. Therefore, appropriate credits for operator action need to be included as part of a "best estimate" analysis because there is no way to establish *a priori* if a particular transient will make a large contribution to the total risk. Nonetheless, the results of our analyses demonstrate that these operator action credits have a small ***overall effect*** on a plant's ***total TWCF***, for reasons detailed below.
    - Medium- and Large-Diameter Primary Side Pipe Breaks: No operator actions are modeled for any break diameter because, for these events, the safety injection systems do not fully refill the upper regions of the reactor coolant system (RCS). Consequently, operators would never take action to shut off the pumps.
    - Stuck-Open Primary Side Valves that May Later Reclose: Reasonable and appropriate credit for operator actions (throttling of the high-pressure injection (HPI) system) has been included in the PRA model. However, these credits have a small influence on the estimated values of vessel failure probability attributable to transients caused by a stuck-open valve in the primary pressure circuit (SO-1 transients) because the credited operator actions only prevent repressurization when SO-1 transients initiate from Hot Zero Power (HZP) conditions and when the operators act promptly (within 1 minute) to throttle the HPI. Complete removal of operator action credits from the model only slightly increases the total risk associated with SO-1 transients.
    - Main Steam Line Breaks: For the overwhelming majority of transients caused by a main steam line break (MSLB), vessel failure is predicted to occur between 10 and 15 minutes after transient initiation because the thermal stresses associated with the rapid cooldown reach their maximum within this timeframe. Thus, all of the long-term effects (isolation of feedwater flow, timing of HPSI control) that can be influenced by operator actions have no effect on vessel failure probability because such factors influence the progression of the transient *after failure has occurred* (if it occurs at all). Only factors affecting the initial cooling rate (i.e., plant power level at time of transient initiation, break location inside or outside of containment) can influence the conditional probability of through-wall cracking (CPTWC), and operator actions do not influence such factors in any way.
- ***Because the severity of the most significant transients in the dominant transient classes is controlled by factors that are common to PWRs in general, the TWCF results presented herein can be used with confidence to develop revised PTS screening criteria that apply to the entire fleet of operating PWRs.***
    - Medium- and Large-Diameter Primary Side Pipe Breaks: For these break diameters, the fluid in the primary cools faster than the wall of the RPV. In this situation, ***only*** the thermal conductivity of the steel and the thickness of the RPV wall control the thermal stresses and, thus, the severity of the fracture challenge. Perturbations in the fluid cooldown rate controlled by break diameter, break location, and season of the year do not play a role. Thermal conductivity is a physical property,

so it is very consistent for all RPV steels, and the thicknesses of the three RPVs analyzed are typical of PWRs. Consequently, the TWCF contribution of medium- to large-diameter primary side pipe breaks is expected to be consistent from plant-to-plant and can be well represented for all PWRs by the analyses reported herein.

- Stuck-Open Primary Side Valves that May Later Reclose: A major contributor to the risk-significance of SO-1 transients is the return to full system pressure once the valve recloses. The operating and safety relief valve pressures of all PWRs are similar. Additionally, as previously noted, operator action credits only slightly affect the total risk associated with this transient class.
- Main Steam Line Breaks: Since MSLBs fail early (within 10–15 minutes after transient initiation), only factors affecting the initial cooling rate can have any influence on the CPTWC values. These factors, which include the plant power level at event initiation and the location of the break (inside or outside of containment), are not influenced by operator actions in any way.

- ***Sensitivity studies performed on the TH and PFM models to investigate the effect of credible model variations on the predicted TWCF values revealed no effects significant enough to recommend changes to the baseline RELAP and FAVOR models, or to recommend cautions regarding the robustness of those models.***
- ***An investigation of design and operational characteristics for five additional PWRs revealed no differences in sequence progression, sequence frequency, or plant thermal-hydraulic response significant enough to call into question the applicability of the TWCF results from the three detailed plant analyses to PWRs in general.***
- ***An investigation of potential external initiating events (e.g., fires, earthquakes, floods) revealed that the contribution of those events to the total TWCF can be regarded as negligible.***

Annual Limit on TWCF

- ***The current guidance provided by Regulatory Guide 1.174 [RG 1.174] for large early release is appropriately applied to setting an acceptable annual TWCF limit of $1x10^{-6}$ events/year.***
  - While many post-PTS accident progressions led only to core damage (which suggests a TWCF limit of $1x10^{-5}$ events/year limit in accordance with Regulatory Guide 1.174), uncertainties in the accident progression analysis led to our recommendation to adopt the more conservative limit of $1x10^{-6}$ events/year based on LERF.

Recommended Revision of the PTS Screening Limits

We recommend using different reference temperature (*RT*) metrics to characterize an RPV's resistance to fractures initiating from different flaws at different locations in the vessel. Specifically, we recommend a reference temperature for flaws occurring along axial weld fusion lines (RT_{AW} or $RT_{AW\text{-}MAX}$), another for flaws occurring in plates or in forgings (RT_{PL} or $RT_{PL\text{-}MAX}$), and a third for flaws occurring along circumferential weld fusion lines (RT_{CW} or $RT_{CW\text{-}MAX}$). In each of these reference temperature pairs, the first metric is a weighted value that accounts for the differences between plants in weld fusion line area or plate volume, while the second metric is a maximum value that can be estimated based only on the information in the NRC's Reactor Vessel Integrity Database (RVID). We also recommend using different *RT* values together to characterize the fracture resistance of the vessel's beltline region, in recognition of the fact that the probability of vessel fracture initiating from different flaw populations varies considerably in response to factors that are both understood and predictable. Correlations between these *RT* metrics and the TWCF attributable to axial weld flaws, plate flaws, and circumferential weld flaws show little plant-to-plant variability because of the general similarity of PTS challenges among plants.

RT-based screening limits were established by setting the total TWCF (i.e., that attributable to axial weld flaws and plate flaws and circumferential weld flaws) equal to the reactor vessel failure frequency acceptance criterion of 1×10^{-6} events per year. The following figures graphically represent these screening limits (for the maximum *RT* metrics), along with an assessment of all operating PWRs relative to these limits. In these figures, the region of the graphs between the red locus and the origin has TWCF values below the 1×10^{-6} acceptance criterion, so these combinations of reference temperatures would be considered acceptable and require no further analysis. By contrast, the region of the graph outside of the red locus has TWCF values above the 1×10^{-6} acceptance criterion, indicating the need for additional analysis or other measures to justify continued plant operation. Clearly, operating PWRs do not closely approach the 1×10^{-6}/year limit. At EOL, at least 70°F, and up to 290°F, (39 to 161°C) separate plate-welded PWRs from the proposed screening limit; this separation between plant-specific values and the proposed screening limit reduces by 10–20°F (5.5 to 11°C) at end of license extension (EOLE, defined as 60 operating years or 48 EFPY). Additionally, no forged plant is anywhere close to the limit of 1×10^{-6} events per year at either EOL or EOLE. This separation of operating plants from the screening limit contrasts markedly with the current situation, where the most embrittled plants are within 1°F (0.5°C) of the screening limit set forth in 10 CFR 50.61. These differences in the "proximity" of operating plants to the current (10 CFR 50.61) and proposed screening limits are illustrated by the bar graph on the next page.

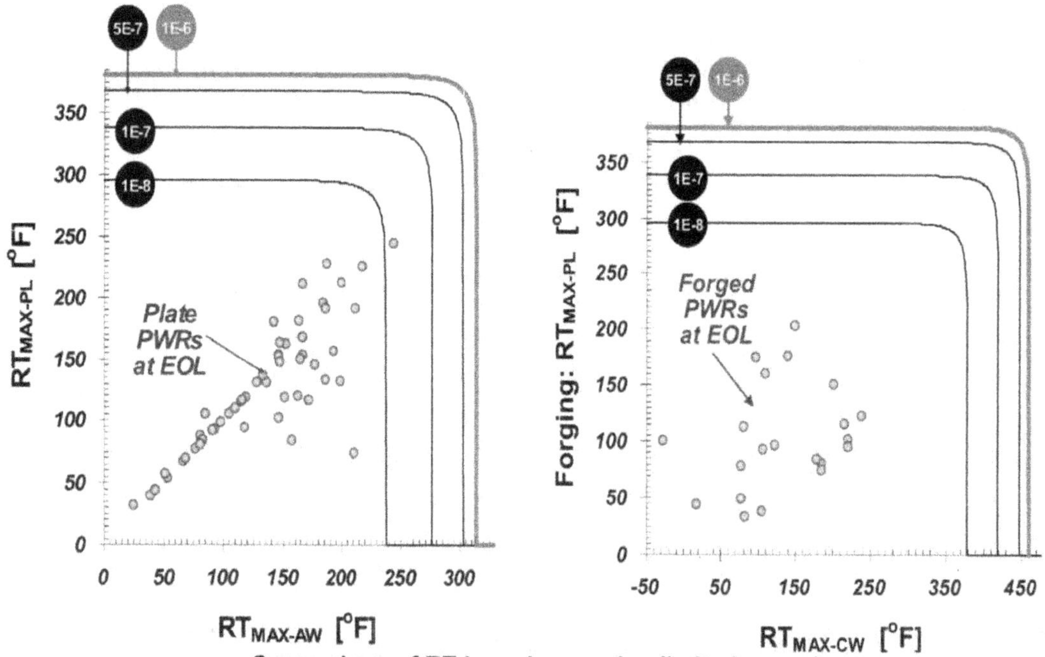

Comparison of RT-based screening limits (curves) with assessment points for operating PWRs at EOL (Left: plate vessels, Right: forged vessels)

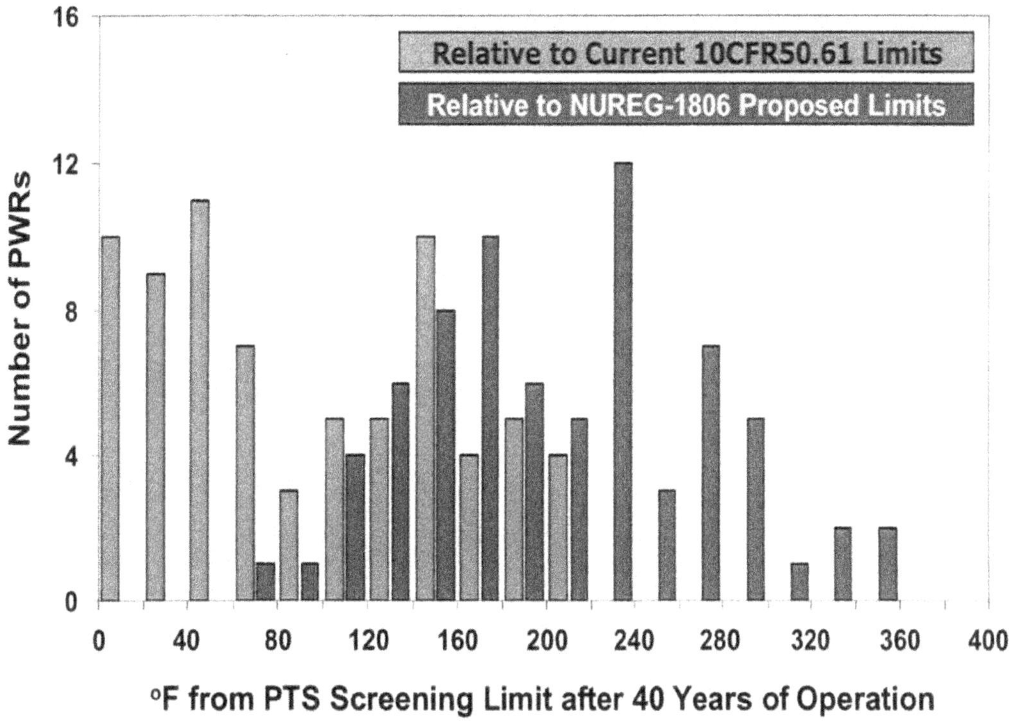

Difference between the proximity of operating PWRs to the current RT_{PTS} screening limits and to the screening limits proposed based on the work presented in this report.

These *RT*-based screening limits (and similar limits described in the text for application to weighted *RT* values) apply to PWRs in general, subject only to the following provisos:

- When assessing a forged vessel where the forging has a very high reference temperature (RT_{PL} above 225°F (107°C)) ***and*** the forging is believed to be susceptible to subclad cracking, a plant-specific analysis of the TWCF produced by the subclad cracks should be performed. However, no forging is projected to reach this level of embrittlement, even at EOLE.
- When assessing an RPV having a wall thickness of 7-in. (18-cm) or less (7 vessels), the proposed *RT* limits are conservative.
- When assessing an RPV having a wall thickness of 11-in. (28-cm) or greater, the proposed *RT* limits may be nonconservative. For the three plants meeting this criterion, either the *RT* limits would need to be reduced or known conservatisms in the current analysis would have to be removed to demonstrate compliance with the TWCF limit of 1×10^{-6} event/year. However, because these three plants are Units 1, 2, and 3 of the Palo Verde Nuclear Generating Station, which have vessels with very low embrittlement projected at EOL and EOLE, there is little practical need for such plant-specific analysis.

Aside from relying on different *RT* metrics than 10 CFR 50.61, this proposed revision of the PTS screening limit differs from the current screening limit in the absence of a "margin term." Use of a margin term is appropriate to account (at least approximately) for factors that occur in application but were not considered in the analysis upon which the screening limit is based. For example, the 10 CFR 50.61 margin term accounts for uncertainty in copper, nickel, and initial RT_{NDT}. However, our model explicitly considers uncertainty in all of these variables, and represents these uncertainties as being larger (a conservative representation) than would be appropriate in any plant-specific application of the proposed screening limit. Consequently, use of the 10 CFR 50.61 margin term with the new screening limits is inappropriate. In general, the following additional reasons suggest that use of *any* margin term with the proposed screening limits is inappropriate:

(1) The *TWCF* values used to establish the screening limit represent 90th percentile values or greater.

(2) The results from our three plant-specific analyses apply to PWRs *in general*, as demonstrated in Chapters 8 and 9 of this report.

(3) Certain aspects of our modeling cannot reasonably be represented as "best estimates." On balance, there is a conservative bias to these non-best-estimate aspects of our analysis because residual conservatisms in the model far outweigh residual nonconservatisms.

Abbreviations

¼-T FLAW	Surface-breaking flaw defined by ASME Boiler and Pressure Vessel Code as having a depth equal to one-quarter of the vessel wall thickness and a length equal to six times the flaw depth
1D	One-Dimensional
ABAQUS	Commercial finite element code developed by Hibbett, Karlsson, and Sorenson in Pawtucket, Rhode Island
ACRS	Advisory Committee on Reactor Safety (NRC)
ADV	Atmospheric Dump Valve
AFW	Auxiliary Feedwater
APET	Accident Progression Event Tree
APEX	Advanced Plant Experiment
ASME	American Society of Mechanical Engineers
ASTM	American Society for Testing and Materials
ATWS	Anticipated Transient without Scram
B&W	Babcock and Wilcox
BWOG	Babcock and Wilcox Owners' Group
BCC	Body-Centered Cubic
BWR	Boiling-Water Reactor
CDF	Core Damage Frequency
CE	Combustion Engineering
CEOG	Combustion Engineering Owners' Group
CFD	Computational Fluid Dynamics
CL	Cold Leg
CFR	*Code of Federal Regulations*
CFT	Core Flood Tank
CPI	Conditional Probability of Crack Initiation
CPTWC	Conditional Probability of Through-Wall Cracking
CSAU	Code Scaling, Applicability, and Uncertainty Methodology
CSNI	Committee on the Safety of Nuclear Installations
CST	Condensate Storage Tank
CVN	Charpy V-Notch
ECC	Emergency Core Cooling
ECCS	Emergency Core Cooling System
EFPY	Effective Full-Power Years
EFW	Emergency Feedwater
EOL	End of License (40 operating years, 32 EFPY)
EOLE	End of License Extension (60 operating years, 48 EFPY)

EPRI	Electric Power Research Institute
ESFAS	Engineered Safety Features Actuation System
F&B	Feed-and-Bleed
FAVOR	Fracture Analysis of Vessels, Oak Ridge
FCI	Frequency of Crack Initiation
GMAW	Gas Metal Arc Weld
H2TS	Hierarchical, Two-Tiered Scaling
HCLPF	High Confidence of Low Probability of Failure
HEP	Human Error Probability
HFE	Human Failure Event
HPI	High-Pressure Injection
HPSI	High-Pressure Safety Injection
HRA	Human Reliability Analysis
HSSI	Heavy Section Steel Irradiation (Project)
HZP	Hot Zero Power
IAEA	International Atomic Energy Agency
ID	Inner Diameter
IPE	Individual Plant Examination
IPEEE	Individual Plant Examination of External Events
IPTS	Integrated Pressurized Thermal Shock
ISLOCA	Interfacing Systems Loss-of-Coolant Accident
ITV	Intermediate Test Vessel
IVO	Imatran Voima Oy
LAS	Low-Alloy Steel
LBLOCA	Large-Break Loss-of-Coolant Accident (pipe diameters above ~8-in. (~20-cm))
LEFM	Linear Elastic Fracture Mechanics
LER	Licensee Event Report
LERF	Large Early Release Frequency
LOCA	Loss-of-Coolant Accident
LOF	Lack of Inter-Run Fusion
LOFT	Loss-of-Fluid Test facility
LPI	Low-Pressure Injection
LPSI	Low-Pressure Safety Injection
MBLOCA	Medium-Break Loss-of-Coolant Accident (pipe diameters of ~4 to 8-in. (~10 to 20-cm))
MFIV	Main Feedwater Isolation Valve
MFW	Main Feedwater
MIST	Multi-loop Integral System Test
MRJ	Materials Reliability Project
MSIV	Main Steam Isolation Valve
MSLB	Main Steam Line Break

NDT	Nil-Ductility Temperature
NEA	Nuclear Energy Agency (OECD)
NRC	U.S. Nuclear Regulatory Commission
NRR	Office of Nuclear Reactor Regulation (NRC)
NUREG/CR	NRC Technical Report Designator (Contractor-prepared Report published by the U.S. Nuclear Regulatory Commission)
OD	Outer Diameter
OECD	Organization for Economic Cooperation and Development
ORNL	Oak Ridge National Laboratory
PFM	Probabilistic Fracture Mechanics
PIRT	Phenomena Identification and Ranking Table
PNNL	Pacific Northwest National Laboratories
PORV	Power-Operated Relief Valve
Ppb	Parts per Billion
PRA	Probabilistic Risk Assessment
PRODIGAL	Probability of Defect Initiation and Growth Analysis
PTS	Pressurized Thermal Shock
PTSE	Pressurized Thermal Shock Experiment
PVRUF	Pressure Vessel Research Users' Facility
PWR	Pressurized-Water Reactor
QHO	Quantitative Health Objective, as defined by the Commission's Safety Goal Policy Statement [NRC FR 86]
RCP	Reactor Coolant Pump
RCS	Reactor Coolant System
RELAP	Reactor Leak and Power excursion code
REMIX	a computer program used to determine the temperature of a plume in the downcomer when the flow in the loops is stagnant
RES	Office of Nuclear Regulatory Research (NRC)
RG	Regulatory Guide
RLE	Review-Level Earthquake
ROSA	Rig of Safety Assessment
RPS	Reactor Protection System
RPV	Reactor Pressure Vessel
RT	Reference Temperature
RVFF	Reactor Vessel Failure Frequency
RVID	Reactor Vessel Integrity Database
RWST	Refueling Water Storage Tank
SAPHIRE	Systems Analysis Programs for Hands-on Integrated Reliability Evaluations
SAW	Submerged Arc Weld
SBLOCA	Small-Break Loss-of-Coolant Accident (pipe diameters below ~4-in. (~10-cm))
SCC	Stress Corrosion Cracking
SECY	Secretary of the (U.S. Nuclear Regulatory) Commission

SEMISCALE	a 1:1705 scaled experimental facility that simulates the primary system of a 4-loop PWR plant
SG	Steam Generator
SGTR	Steam Generator Tube Rupture
SIAS	Safety Injection Actuation Signal
SIT	Safety Injection Tank
SMAW	Submerged Metal Arc Weld
SO-1	Stuck-open valve in the primary pressure circuit
SO-2	Stuck-open valve in the secondary pressure circuit
SQA	Software Quality Assurance
SRM	Staff Requirements Memorandum
SRV	Safety/Relief Valve
SSC	System, Structure, or Component
SSE	Safe-Shutdown Earthquake
SSRV	Secondary System Relief Valve
TBV	Turbine Bypass Valve
TH	Thermal-Hydraulics
TMI	Three Mile Island
TSE	Thermal Shock Experiment
TWCF	Through-Wall Cracking Frequency
UMD	University of Maryland
UPTF	Upper Plenum Test Facility
USE	Charpy V-Notch Upper-Shelf Energy
V&V	Verification and Validation
VCIF	Vessel Crack Initiation Frequency
(W)	Westinghouse
WOG	Westinghouse Owners' Group
WPS	Warm Pre-Stress

Nomenclature

Symbols Used in Thermal-Hydraulics

α	thermal diffusivity, m^2/s
β	bulk coefficient of expansion, 1/C
μ	viscosity, kg/m-s
ν	kinematic viscosity, m^2/s
ρ	density, kg/m^3
σ	stress, kg/s^2
τ	characteristic time
C_p	heat capacity, m^2/s^2-C
g	gravitational acceleration, m/s^2
Gr	Grashof Number
h	convective heat transfer coefficient, W/m^2-C
D	diameter, m
J	joules, $kg\text{-}m^2/s^2$
k	conductivity, W/m-C
l	length, m
Nu	Nusselt Number
Pr	Prandtl Number
P	pressure, $kg/m\text{-}s^2$
q	heat flux, W/m^2
Re	Reynolds Number
Ri	Richardson Number
s	seconds
t	thickness, m
t	time, s
u	velocity, m/s
T	temperature, C
W	watts, $kg\text{-}m^2/s^3$

Symbols Used in Fracture Mechanics

$2a$	Flaw depth measured through the vessel wall thickness
$2c$	Flaw length measured parallel to the axial or circumferential direction of the vessel
Cu	Copper content, weight%
J_{Ic}	A fracture toughness measure defined by ASTM E1820, which quantifies the resistance of metals to crack initiation by the initiation, growth, and coalescence of microvoids
$J\text{-}R$	A fracture toughness measure defined by ASTM E1820, which quantifies the resistance of metals to ductile tearing
K_{Jc}	A fracture toughness measure defined by ASTM E1921, which quantifies the resistance of metals to crack initiation by cleavage mechanisms
K_{Ia}	A fracture toughness measure defined by ASTM E1221, which quantifies the ability of metals to arrest (stop) a running cleavage crack
K_{Ic}	A fracture toughness measure defined by ASTM E399, which quantifies the resistance of metals to crack initiation under plane strain conditions
$K_{Ic(min)}$	The minimum K_{Ic} fracture toughness possible at a particular temperature
$K_{APPLIED}$	Linear elastic crack driving force
$\mathcal{L}$	For a buried defect, distance from the wetted clad surface on the vessel ID to the inner crack tip
ℓ	The length of the fusion line of an axial weld
Ni	Nickel content, weight%
P	Phosphorus content, weight%
RT_{AW}	A fracture toughness reference temperature, which characterizes the RPV's resistance to fractures initiating from flaws found along the axial weld fusion lines. It corresponds to the maximum RT_{NDT} of the plates/welds that lie to either side of the weld fusion lines, and is weighted to account for differences in weld fusion line length (and, therefore, number of simulated flaws) between vessel courses.
RT_{PL}	A fracture toughness reference temperature, which characterizes the RPV's resistance to fractures initiating from flaws found in plates that are not associated with welds. It corresponds to the maximum RT_{NDT} occurring anywhere in the plate.
RT_{CW}	A fracture toughness reference temperature, which characterizes the RPV's resistance to fractures initiating from flaws found along the circumferential weld fusion lines. It corresponds to the maximum RT_{NDT} of the plates/welds that lie to either side of the weld fusion lines.
RT_{NDT}	Transition fracture toughness reference temperature defined by ASME NB-2331
$RT_{NDT(u)}$	Unirradiated value of RT_{NDT}
RT_{PTS}	RT_{NDT} projected end of license to account for the effects of irradiation (defined in 10 CFR 50.61)
t_{WALL}	Vessel wall thickness
t_{CLAD}	Stainless steel cladding thickness

T_{30}	The temperature at which the mean CVN energy is 30 ft-lbs (41J)
$T_{35/50}$	Charpy V-notch energy transition temperature defined as the temperature at which the CVN energy is at least 50 ft-lbs (68J) and the lateral expansion of the specimen is at least 0.035-in. (0.89-mm) [See the definition on page 2-7]
T_{NDT}	Nil-ductility temperature defined by ASTM E-208
ΔT_{30}	The shift in the CVN 30 ft-lb (41J) transition temperature produced by radiation damage
σ_{flow}	Flow strength, average of tensile yield and tensile ultimate strength
ϕt	Fluence

Glossary

Terms Used in Probabilistic Risk Assessment

Abnormal operating procedure	A procedure (i.e., list of actions) used to address unique or special plant circumstances identified while using emergency operating procedures (EOPs). These abnormal operating procedures are usually called by EOPs, but may be indicated directly by some plant conditions.
Accident progression event tree	The event tree used to model the part of the accident sequence that follows the onset of core damage, including containment response to severe accident conditions, equipment availability, and operator performance.
Binning	The process of taking a large number of sequences and combining then into a smaller number of groups, that are expected to have similar characteristics (e.g., TH conditions), to allow effective utilization of limited resources.
Core damage	Uncovery and heatup of the reactor core to the point at which prolonged oxidation and severe fuel damage is anticipated and involving enough of the core to cause a significant release.
Dominant scenario	An accident sequence (scenario) that is usually represented by the top 10 or 20 events or groups of events modeled in a PRA, which accounts for a large fraction of the specified end state.
Emergency operating procedure	The primary procedure (i.e., list of actions) used to respond to a plant disturbance resulting from an initiating event.
Event tree	A logic diagram that begins with an initiating event or condition and progresses through a series of branches that represent expected system or operator performance that either succeeds or fails and arrives at either a successful or failed end state.
Fault tree	A deductive logic diagram that depicts how a particular undesired event can occur as a logical combination of other undesired events.
Large Early Release	The rapid, unmitigated release of airborne fission products from the containment to the environment occurring before the effective implementation of offsite emergency response and protective actions, such that there is a potential for early health effects.
Latin Hypercube sampling	A stratified sampling technique, in which the random variable distributions are divided into equal probability intervals, and probabilities are then randomly selected from within each interval.
Mitigating equipment	Systems or components, used to respond to an initiating event, of which successful operation prevents the occurrence of an undesired event or state.
Pre-initiator human failure event	Human failure events that represent the impact of human errors committed during actions performed prior to the initiation of an accident (e.g., during maintenance or the use of calibration procedures).
Post-initiator human failure event	Human failure events that represent the impact of human errors committed during actions performed in response to an accident initiator.

Prompt fatality	A fatality that results from substantial radiation exposures incurred during short time periods (usually within weeks, though up to 1 year for pulmonary effects).
PTS bin	A group of sequences that are expected to have similar TH characteristics and are represented by one unique set of TH characteristics during a FAVOR calculation.
Risk-informed	An approach to analyzing and evaluating activities, which bases decisions on the results of traditional engineering evaluations, supported by insights derived from the use of PRA methods.
Scenario	See Sequence.
Screening	The process of eliminating items from further consideration based on their negligible contribution to the probability of an undesired end state or its consequences.
Sequence	A representation in terms of an initiating event followed by a sequence of failures or successes of events (i.e., system, function, or operator performance) that can lead to undesired consequences, with a specified end state (e.g., potential for PTS).

Terms Used in Thermal-Hydraulics

Blowdown	Rapid depressurization of a system in response to a break. .
Break flow	Flow of water (liquid and vapor) out a pipe break or a valve.
Break energy	Energy content of the fluid flow out a break.
Bottom-up	To break up a complex system into its subsystems, and then break up each subsystem into its components, examine individual local phenomena and processes that most affect each component, and build up the total complex system from these individual pieces (like manufacturing a car).
Coast down	Time required for a pump to stop rotating once power is shut off due to inertia.
Decay heat	Heat generated from radioactive decay of fission products.
Enthalpy	Sum of internal energy and volume multiplied by pressure.
Flash	Change of phase from saturated liquid to vapor resulting from decrease in pressure.
Flow quality	Mass fraction of flow stream that is steam. Higher quality flow would have a high mass fraction of steam.
Forced flow	Flow driven by a pump.
Inventory	Mass of water.
Loop flow	Mass flow rate of coolant in a circuit.
Makeup water	Water reservoir available for inventory control.
Natural circulation	Flow driven by buoyancy (gravity).
Pressure drop	Change in pressure due to conversion of mechanical energy to internal energy.
Protection system	Electrical controls to actuate engineering safety features.
Quality	Mass fraction of steam in a two-phase steam-water mixture.

Saturation temperature	A temperature corresponding to phase change from liquid to vapor.
Sensible heat	The product of specific heat and temperature change of subcooled liquid.
Subcooled	A system is *subcooled* if it exists entirely in a liquid state. The *degree of subcooling* is the number of degrees that the temperature of the system would have to be raised to cause boiling.
Throttled	Operation of a control valve to regulate flow.
Top-down	To characterize a complex system by establishing the governing behavior, or phenomenon, that is most important, and then proceed from that starting point to successive lower levels, by identifying the processes that have the greatest influence on the top-level phenomenon.
Trip	A "trip" occurs when a breaker opens in response to its trip mechanism (an arm that holds the breaker closed moves to allow the breaker to open). When a reactor trips, all of the breakers that provide power to the rod control system open, causing the rods to be inserted in the core and stopping the nuclear reaction. When a pump trips, the breaker opens, thereby disconnecting power and causing the pump to stop.
Water solid	A situation in which there is no steam in the system (i.e., it is all liquid). A "water solid" system is *subcooled.*

Terms Used in Fracture Mechanics

Brittle	Fracture occurring without noticeable macroscopic plastic deformation (stretching) of the material.
Cleavage fracture	Microscopically, cleavage is a fracture mode that occurs preferentially along certain atomic planes through the grains of the material. Cleavage can only occur in ferritic steels (i.e., steels having a body-centered cubic lattice structure). Macroscopically, cleavage fracture is often called "brittle" fracture because little noticeable plastic deformation (stretching) of the material occurs. (Note, however, that plastic flow at the micro-scale is a necessary precursor to cleavage.) Macroscopically, cleavage fracture is also characterized as being a sudden event, with cracks of very large dimensions developing over durations measured in fractional seconds. A useful, although inexact, analogue for cleavage fracture in common experience is the breaking of glass.
Ductile fracture	Microscopically, ductile fracture occurs through the initiation, growth, and eventual coalescence of micro-voids in the material into a macroscopic crack. These micro-voids tend to initiate at local heterogeneities in the material (e.g., inclusions, carbides, clusters of dislocations). Macroscopically, ductile fracture is associated with considerable plastic deformation (stretching) of the material. Relative to cleavage fracture, ductile fracture occurs very slowly, with crack growth rates measured in seconds rather than in micro-seconds (for cleavage).
Fracture toughness	A general term referring to a material's resistance to fracture. The term may be modified to refer to fractures by different mechanisms: <u>Arrest fracture toughness</u> measures a material's ability to stop a running cleavage crack. <u>Cleavage fracture toughness</u> measures a material's ability to resist crack initiation in cleavage. <u>Ductile fracture toughness</u> measures a material's ability to resist crack initiation attributable to ductile mechanisms on the upper shelf.

Lower shelf	At low temperatures, the toughness behavior of steels occurs by transgranular cleavage and is said to be on the lower shelf. On the lower shelf, a fracture is unstable, and is often referred to as a "brittle" fracture.
Reference temperature	A characteristic temperature used to locate the transition curve of a ferritic steel on the temperature axis.
Transition (or transition curve)	Between lower shelf and upper shelf temperatures, the fracture behavior of a ferritic material is said to be in "transition." At low temperatures in transition, fracture occurs by cleavage. As temperature increases through the transition regime, fracture occurs by ductile crack initiation and growth, a process which is terminated by cleavage. At still higher temperatures, cleavage cannot occur, and upper shelf conditions exist.
Upper shelf	At high temperatures, the toughness behavior of steels occurs by ductile mechanisms (micro-void initiation, growth, and coalescence) and is said to be on the upper shelf. On the upper shelf, afracture is stable and dissipates considerable amounts of energy.

Terms Used in Uncertainty Analysis

Aleatory	Aleatory uncertainties arise as a result of the randomness inherent in a physical or human process. Consequently, aleatory uncertainties are fundamentally irreducible. If the uncertainty in a variable is characterized as being aleatory, the entire distribution of the variable is carried through each simulation run.
Epistemic	Epistemic uncertainties are caused by limitations in our current state of knowledge (or understanding) of a given process. Epistemic uncertainties can, in principle, be reduced by an increased state of knowledge. If the uncertainty in a variable is characterized as being epistemic in a probabilistic simulation, individual values of the variable are randomly selected from a distribution and propagated through the calculation. This procedure models the understanding that the "correct" value of the variable is knowable, at least in principal. Thus, for epistemic uncertainties, individual simulation runs are deterministic, while the totality of all simulation runs captures the uncertainty characteristic of the epistemic variable.

Appendix A – Master Transient List and FAVOR 04.1 Results Summary

Table A.1. Transient descriptions and FAVOR 04.1 results for medium- and large-diameter pipe break (LOCA) transients

#	Dia [in]	System Failure	IEF	Percent Contribution to Total Frequency of Crack Initiation (FCI)				Percent Contribution to Total Through Wall Cracking Frequency (TWCF)				Mean CPI				Mean CPTWC			
				32	60	Ext-A	Ext-B	32	60	Ext-A	Ext-B	32	60	Ext-A	Ext-B	32	60	Ext-A	Ext-B
Beaver Valley Unit 1																			
9	16	2.54-cm [16-in.] hot leg break	6.99E-06	5.0	5.8	6.5	7.6	1.5	3.3	10.5	10.2	6.40E-04	3.02E-03	1.21E-02	7.44E-02	8.63E-07	1.43E-05	1.58E-04	2.66E-03
7	8	2.54-cm [8-in.] surge line break	2.11E-05	14.4	15.7	17.2	17.4	17.2	28.6	19.4	14.9	6.16E-04	2.58E-03	9.06E-03	4.66E-02	2.01E-06	2.08E-05	1.44E-04	1.52E-03
117	5.7	14.366-cm [5.657-in.] cold leg break, summer conditions (HHSI, LHSI temp = 55°F, Accumulator Temp = 105°F)	2.11E-05	0.2	0.2	0.2	0.2	0.0	0.0	0.0	0.0	1.27E-05	4.66E-05	1.36E-04	6.45E-04	8.29E-10	2.29E-08	9.46E-08	1.14E-06
116	5.7	14.366-cm [5.657-in.] cold leg break with break area increased 30%	1.81E-05	0.0	0.0	0.1	0.1	0.0	0.0	0.0	0.0	1.27E-06	9.46E-06	4.85E-05	4.33E-04	2.15E-10	1.27E-08	6.23E-08	2.16E-06
56	4	10.16-cm [4.0-in.] surge line break (see Note at end of table)	1.23E-04	79.4	77.1	74.2	68.9	13.2	30.8	51.4	43.2	7.43E-04	2.84E-03	9.36E-03	4.46E-02	8.76E-07	1.07E-05	8.29E-05	9.82E-04

#	Dia [in]	System Failure	IEF	Percent Contribution to Total Frequency of Crack Initiation (FCI)				Percent Contribution to Total Through Wall Cracking Frequency (TWCF)				Mean CPI				Mean CPTWC			
				32	60	Ext-A	Ext-B	32	60	Ext-A	Ext-B	32	60	Ext-A	Ext-B	32	60	Ext-A	Ext-B
Oconee Unit 1																			
156	16	40.64-cm [16-in.] hot leg break. ECC suction switch to the containment sump included in the analysis.	7.03E-06	6.8	13.2	26.9	28.5	0.0	0.1	4.6	6.3	1.63E-06	1.74E-05	2.81E-03	1.55E-02	2.88E-10	6.82E-09	6.21E-06	7.83E-05
164	8	20.32-cm [8 in.] surge line break. ECC suction switch to the containment sump included in the analysis.	2.12E-05	41.2	38.9	36.1	39.7	0.1	1.2	17.1	29.6	1.20E-06	1.09E-05	1.29E-03	7.06E-03	7.78E-11	9.05E-09	6.85E-06	8.05E-05
160	5.7	14.37-cm [5.656-in.] surge line break. ECC suction switch to the containment sump included in the analysis.	1.82E-05	33.3	39.8	32.4	26.3	0.1	0.6	24.6	32.8	2.77E-06	1.97E-05	1.32E-03	5.84E-03	1.36E-09	4.01E-08	1.33E-05	1.21E-04
172	4	10.16-cm [4-in.] cold leg break. ECC suction switch to the containment sump included in the analysis.	1.06E-04	0.0	0.1	0.8	1.5	-	-	0.5	1.2	4.90E-10	8.71E-09	6.86E-06	6.29E-05	-	-	9.55E-08	2.27E-06
Palisades																			
40	16	40.64-cm (16-in.) hot leg break. Containment sump recirculation included in the analysis.	3.22E-05	65.3	63.7	58.2	58.7	14.4	20.5	27.7	30.5	7.60E-04	1.76E-03	1.03E-02	6.42E-02	1.36E-05	7.11E-05	1.21E-03	9.57E-03
62	8	20.32-cm (8-in.) cold leg break. Winter conditions assumed (HPI and LPI injection temp = 40 F, Accumulator temp = 60 F)	7.07E-06	6.9	6.6	6.6	6.9	2.1	3.2	4.2	4.7	3.73E-04	8.85E-04	5.31E-03	3.37E-02	1.08E-05	5.50E-05	8.57E-04	6.59E-03
63	5.7	14.37-cm (5.656-in.) cold leg break. Winter conditions assumed (HPI and LPI injection temp = 40 F, Accumulator temp = 60 F)	6.06E-06	2.0	2.0	2.3	2.3	0.8	1.4	2.0	2.1	1.41E-04	3.37E-04	2.08E-03	1.30E-02	5.00E-06	2.72E-05	4.53E-04	3.30E-03

#	Dia [in]	System Failure	IEF	Percent Contribution to Total Frequency of Crack Initiation (FCI)				Percent Contribution to Total Through Wall Cracking Frequency (TWCF)				Mean CPI				Mean CPTWC			
				32	60	Ext-A	Ext-B	32	60	Ext-A	Ext-B	32	60	Ext-A	Ext-B	32	60	Ext-A	Ext-B
58	4	10.16-cm (4-in.) cold leg break. Winter conditions assumed (HPI and LPI injection temp = 40 F, Accumulator temp = 60 F)	2.66E-04	11.5	11.7	11.8	11.3	4.9	8.8	12.6	12.4	2.01E-05	4.79E-05	3.05E-04	1.78E-03	8.11E-07	4.47E-06	7.88E-05	5.34E-04
64	4	10.16-cm (4-in.) surge line break. Summer conditions assumed (HPI and LPI injection temp = 100 F, Accumulator temp = 90 F)	7.07E-06	4.0	3.8	3.4	3.3	2.2	3.5	3.7	3.4	2.35E-04	5.28E-04	2.95E-03	1.68E-02	1.21E-05	6.03E-05	7.92E-04	4.88E-03
59	4	10.16-cm (4-in.) cold leg break. Summer conditions assumed (HPI and LPI injection temp = 100 F, Accumulator temp = 90 F)	2.09E-04	0.6	0.8	1.0	1.2	0.1	0.3	0.7	1.0	1.34E-06	3.94E-06	3.32E-05	2.15E-04	2.00E-08	1.72E-07	5.73E-06	5.37E-05

<u>Note</u>: There are no operator actions for any of these transients, and all transients initiate from full power conditions **<u>except for</u>** Beaver Valley 56, which initiates from hot zero power conditions. However, Beaver Valley 56 is used to represent full power conditions in this analysis.

Table A.2. Transient descriptions and FAVOR 04.1 results for small-diameter pipe break (LOCA) transients

#	Dia [in]	System Failure	IEF	Percent Contribution to Total Frequency of Crack Initiation (FCI)				Percent Contribution to Total Through Wall Cracking Frequency (TWCF)				Mean CPI				Mean CPTWC			
				32	60	Ext-A	Ext-B	32	60	Ext-A	Ext-B	32	60	Ext-A	Ext-B	32	60	Ext-A	Ext-B
Beaver Valley Unit 1																			
114	2.8	7.18-cm [2.828-in.] surge line break, summer conditions (HHSI, LHSI temp = 55°F, Accumulator Temp = 105°F), heat transfer coefficient increased 30% (modeled by increasing heat transfer surface area by 30% in passive heat structures).	9.76E-05	0.24	0.35	0.54	0.70	0.00	0.01	0.03	0.08	2.55E-06	1.52E-05	6.70E-05	4.57E-04	6.63E-11	8.69E-09	7.25E-08	2.45E-06
115	2.8	7.18-cm [2.828-in.] cold leg break	9.76E-05	-	-	-	-	-	-	-	-	-	-	-	-	-	-	-	-
3	2	5.08-cm [2-in.] surge line break	9.76E-05	0.00	0.02	0.03	0.08	-	0.00	0.00	0.01	3.81E-08	6.85E-07	4.01E-06	5.79E-05	-	2.28E-10	2.18E-09	3.57E-07
2	1.4	3.59-cm [1.414-in.] surge line break	1.23E-04	-	-	-	-	-	-	-	-	-	-	-	-	-	-	-	-
Oconee Unit 1																			
154	3.4	8.53-cm [3.36-in.] surge line break [Break flow area reduced by 30% from 10.16-cm [4-in.] break]. Vent valves do not function. ECC suction switch to the containment sump included in the analysis.	1.34E-04	0.00	0.00	0.15	0.36	-	-	0.10	0.79	0.00E+00	1.56E-10	1.18E-06	1.44E-05	-	-	1.44E-08	1.08E-06
178	3.4	8.53-cm [3.36-in.] surge line break [Break flow area reduced by 30% from 10.16-cm [4-in.] break]. Vent valves do not function. ECC suction switch to the containment sump included in the analysis.	2.12E-05	0.00	0.00	0.04	0.07	-	-	0.03	0.21	0.00E+00	1.56E-10	1.18E-06	1.44E-05	-	-	1.44E-08	1.08E-06
141	3.2	8.19-cm [3.22-in.] surge line break [Break flow area increased by 30% from 7.18-cm [2.828-in.] break].	1.06E-04	0.64	1.40	1.71	1.45	0.00	0.06	1.92	2.66	2.00E-08	2.05E-07	1.54E-05	6.44E-05	2.40E-13	7.57E-10	3.77E-07	4.21E-06

#	Dia [in]	System Failure	IEF	Percent Contribution to Total Frequency of Crack Initiation (FCI)				Percent Contribution to Total Through Wall Cracking Frequency (TWCF)				Mean CPI				Mean CPTWC			
				32	60	Ext-A	Ext-B	32	60	Ext-A	Ext-B	32	60	Ext-A	Ext-B	32	60	Ext-A	Ext-B
142	2.4	6.01-cm [2.37-in.] surge line break [Break flow area decreased by 30% from 7.18-cm [2.828-in.] break].	1.06E-04	-	-	-	0.00	-	-	-	-	-	-	-	2.24E-09	-	-	-	-
145	1.7	4.34-cm [1.71-in.] surge line break [Break flow area increased by 30% from 3.81-cm [1.5-in.] break]. Winter conditions assumed [HPI, LPI temp = 277 K [40° F] and CFT temp = 294 K [70° F]].	1.34E-04	-	-	-	-	-	-	-	-	-	-	-	-	-	-	-	-
Palisades																			
61	2.8	7.18-cm (2.8-in.) cold leg break. Summer conditions assumed (HPI and LPI injection temp = 100 F, Accumulator temp = 90 F)	2.09E-04	0.01	0.03	0.11	0.16	0.00	0.01	0.04	0.11	2.98E-08	1.89E-07	3.87E-06	3.06E-05	3.57E-10	3.32E-09	3.71E-07	6.55E-06
60	2	5.08-cm (2-in.) surge line break. Winter conditions assumed (HPI and LPI injection temp = 40 F, Accumulator temp = 60 F)	2.09E-04	1.67	1.94	2.60	2.56	0.89	1.69	3.00	3.51	3.82E-06	1.07E-05	8.22E-05	5.02E-04	1.81E-07	1.13E-06	2.41E-05	1.88E-04
2	1.4	3.59-cm (1.414-in.) surge line break. Containment sump recirculation included in the analysis.	2.66E-04	-	-	-	0.00	-	-	-	0.00	-	-	-	1.41E-14	-	-	-	2.82E-15

Note: **There are no operator actions for any of these transients, and all transients initiate from full power conditions.**

Table A.3. Transient descriptions and FAVOR 04.1 results for stuck-open primary valve transients (including value reclosure)

TH #	Transients Including Valve Reclosure System Failure	Operator Action	HZP	IEF	Percent Contribution to Total Frequency of Crack Initiation (FCI)				Percent Contribution to Total Through Wall Cracking Frequency (TWCF)				Mean CPI				Mean CPTWC			
					32	60	Ext-A	Ext-B	32	60	Ext-A	Ext-B	32	60	Ext-A	Ext-B	32	60	Ext-A	Ext-B
Beaver Valley Unit 1																				
126	Reactor/turbine trip w/one stuck-open pressurizer SRV, which recloses at 6,000 s	Operator controls HHSI 10 minutes after allowed.	N	1.87E-04	0.2	0.1	0.0	0.0	23.6	14.0	3.1	0.77	8.24E-07	2.57E-06	2.79E-06	1.40E-05	8.19E-07	2.50E-06	2.50E-06	1.04E-05
60	Reactor/turbine trip w/one stuck-open pressurizer SRV, which recloses at 6,000 s.	None.	N	2.15E-05	0.1	0.1	0.0	0.0	16.8	7.5	1.8	0.41	4.69E-06	1.17E-05	1.39E-05	4.24E-05	4.69E-06	1.17E-05	1.39E-05	4.20E-05
130	Reactor/turbine trip w/one stuck-open pressurizer SRV, which recloses at 3,000 s at HZP	Operator controls HHSI 10 minutes after allowed.	Y	3.09E-05	0.1	0.1	0.0	0.0	16.5	7.1	1.0	0.24	4.14E-06	8.46E-06	6.53E-06	3.69E-05	4.14E-06	8.45E-06	6.37E-06	1.99E-05
97	Reactor/turbine trip w/one stuck-open pressurizer SRV, which recloses at 3,000 s.	None.	Y	3.74E-06	0.0	0.0	0.0	0.0	4.0	2.0	0.6	0.13	7.68E-06	1.82E-05	1.77E-05	6.03E-05	7.68E-06	1.82E-05	1.77E-05	5.27E-05
129	Reactor/turbine trip w/one stuck-open pressurizer SRV, which recloses at 6,000 s at HZP	Operator controls HHSI 10 minutes after allowed.	Y	3.09E-05	0.0	0.0	0.0	0.0	2.9	1.2	0.2	0.03	9.54E-07	1.62E-06	1.76E-06	2.08E-05	9.54E-07	1.61E-06	1.60E-06	1.99E-05
123	Reactor/turbine trip w/two stuck-open pressurizer SRVs, which reclose at 3,000 s at HZP	Operator controls HHSI 10 minutes after allowed.	Y	1.65E-07	0.0	0.0	0.0	0.0	1.2	0.6	0.3	0.09	6.32E-05	1.78E-04	3.90E-04	1.81E-03	5.59E-05	1.50E-04	2.86E-04	1.38E-03
71	Reactor/turbine trip w/one stuck-open pressurizer SRV, which recloses at 6,000 s.	None.	Y	3.74E-06	0.0	0.0	0.0	0.0	1.2	0.5	0.1	0.02	1.96E-06	4.00E-06	3.57E-06	1.94E-05	1.96E-06	4.00E-06	3.49E-06	8.10E-06

TH #	Transients Including Valve Reclosure System Failure	Operator Action	HZP	IEF	Percent Contribution to Total Frequency of Crack Initiation (FCI)				Percent Contribution to Total Through Wall Cracking Frequency (TWCF)				Mean CPI				Mean CPTWC			
					32	60	Ext-A	Ext-B	32	60	Ext-A	Ext-B	32	60	Ext-A	Ext-B	32	60	Ext-A	Ext-B
61	Reactor/turbine trip w/two stuck-open pressurizer SRV, which recloses at 3,000 s.	None.	N	1.79E-06	0.0	0.0	0.0	0.0	0.2	0.3	0.4	0.26	5.67E-06	2.93E-05	1.15E-04	7.15E-04	1.20E-06	8.87E-06	4.14E-05	4.14E-04
69	Reactor/turbine trip w/two stuck-open pressurizer SRVs, which reclose at 3,000 s.	None.	Y	2.09E-08	0.0	0.0	0.0	0.0	0.3	0.2	0.1	0.04	1.32E-04	4.34E-04	1.18E-03	5.68E-03	9.65E-05	3.09E-04	8.31E-04	4.72E-03
120	Reactor/turbine trip w/two stuck-open pressurizer SRVs, which recloses at 6,000 s	Operator controls HHSI 10 minutes after allowed.	N	9.98E-07	0.0	0.0	0.0	0.0	0.5	0.2	0.0	0.01	3.62E-06	1.14E-05	3.17E-05	2.63E-04	3.07E-06	6.71E-06	6.64E-06	2.69E-05
124	Reactor/turbine trip w/two stuck-open pressurizer SRVs, which reclose at 6,000 s at HZP	Operator controls HHSI 10 minutes after allowed.	Y	1.65E-07	0.0	0.0	0.0	0.0	0.1	0.0	0.0	0.01	1.39E-05	5.49E-05	2.00E-04	1.27E-03	4.16E-06	8.91E-06	1.17E-05	1.06E-04
92	Reactor/turbine trip w/two stuck-open pressurizer SRVs, one recloses at 3000 s.	None.	Y	2.13E-07	0.0	0.0	0.0	0.0	0.0	0.0	0.0	0.01	5.15E-05	2.35E-04	8.71E-04	4.82E-03	2.04E-07	1.87E-06	1.15E-05	1.21E-04
93	Reactor/turbine trip w/two stuck-open pressurizer SRVs. One valve recloses at 6000 seconds, while the other valve remains open.	None.	Y	2.13E-07	0.0	0.0	0.0	0.0	0.0	0.0	0.0	0.01	5.15E-05	2.35E-04	8.71E-04	4.82E-03	2.04E-07	1.87E-06	1.15E-05	1.21E-04
70	Reactor/turbine trip w/two stuck-open pressurizer SRVs, which reclose at 6,000 s.	None.	Y	2.09E-08	0.0	0.0	0.0	0.0	0.0	0.0	0.0	0.00	1.35E-05	5.60E-05	2.32E-04	1.82E-03	6.03E-06	1.25E-05	1.78E-05	1.25E-04
66	Reactor/turbine trip w/two stuck-open pressurizer SRVs. One valve recloses at 3000 seconds, while the other valve remains open.	None.	N	1.18E-06	0.0	0.0	0.0	0.0	0.0	0.0	0.0	0.01	5.25E-06	2.80E-05	1.13E-04	7.05E-04	1.42E-08	2.15E-07	1.41E-06	1.83E-05
62	Reactor/turbine trip w/two stuck-open pressurizer SRV, which recloses at 6,000 s.	None.	N	1.08E-07	0.0	0.0	0.0	0.0	0.0	0.0	0.0	0.00	1.20E-06	4.64E-06	1.61E-05	1.57E-04	9.61E-07	2.12E-06	2.14E-06	7.76E-06

TH #	Transients Including Valve Reclosure System Failure	Operator Action	HZP	IEF	Percent Contribution to Total Frequency of Crack Initiation (FCI)				Percent Contribution to Total Through Wall Cracking Frequency (TWCF)				Mean CPI				Mean CPTWC			
					32	60	Ext-A	Ext-B	32	60	Ext-A	Ext-B	32	60	Ext-A	Ext-B	32	60	Ext-A	Ext-B
59	Reactor/turbine trip w/one stuck-open pressurizer SRV, which recloses at 3,000 s.	None.	N	3.46E-04	0.0	0.0	0.0	0.0	0.0	0.0	0.0	0.00	0.00E+00	0.00E+00	4.10E-12	1.97E-07	0.00E+00	0.00E+00	0.00E+00	1.61E-08
67	Reactor/turbine trip w/two stuck-open pressurizer SRVs. One valve recloses at 6000 seconds, while the other valve remains open.	None.	N	1.18E-06	0.0	0.0	0.0	0.0	0.0	0.0	0.0	0.00	2.07E-07	2.21E-06	1.24E-05	1.38E-04	0.00E+00	2.51E-10	4.49E-09	5.39E-07
119	Reactor/turbine trip w/two stuck-open pressurizer SRV, which recloses at 6,000 s	Operator controls HHSI 1 minute after allowed.	N	6.84E-07	0.0	0.0	0.0	0.0	0.0	0.0	0.0	0.00	5.60E-07	4.78E-06	2.57E-05	2.51E-04	1.76E-12	5.42E-09	6.34E-08	2.81E-06
121	Reactor/turbine trip w/two stuck-open pressurizer SRV, which recloses at 3,000 s at HZP	Operator controls HHSI 1 minute after allowed	Y	1.33E-07	0.0	0.0	0.0	0.0	0.0	0.0	0.0	0.00	9.79E-06	4.63E-05	1.91E-04	1.26E-03	6.92E-11	1.90E-08	5.74E-07	2.89E-05
122	Reactor/turbine trip w/two stuck-open pressurizer SRVs, which reclose at 6,000 s at HZP	Operator controls HHSI 1 minute after allowed.	Y	1.33E-07	0.0	0.0	0.0	0.0	0.0	0.0	0.0	0.00	9.79E-06	4.63E-05	1.91E-04	1.26E-03	3.57E-12	7.81E-09	1.14E-07	5.81E-06
125	Reactor/turbine trip w/one stuck-open pressurizer SRV, which recloses at 6,000 s	Operator controls HHSI 1 minute after allowed.	N	1.34E-04	0.0	0.0	0.0	0.0	0.0	0.0	0.0	0.00	2.39E-09	7.20E-08	3.27E-07	4.87E-06	0.00E+00	5.58E-19	7.77E-11	1.16E-08
127	Reactor/turbine trip w/one stuck-open pressurizer SRV, which recloses at 6,000 s at HZP	Operator controls HHSI 1 minute after allowed.	Y	2.59E-05	0.0	0.0	0.0	0.0	0.0	0.0	0.0	0.00	4.37E-18	8.95E-09	1.62E-07	1.75E-05	0.00E+00	0.00E+00	0.00E+00	1.18E-07
128	Reactor/turbine trip w/one stuck-open pressurizer SRV, which recloses at 3,000 s at HZP	Operator controls HHSI 1 minute after allowed.	Y	2.59E-05	0.0	0.0	0.0	0.0	0.0	0.0	0.0	0.00	4.37E-18	8.95E-09	1.62E-07	1.75E-05	0.00E+00	0.00E+00	0.00E+00	1.18E-07

Oconee Unit 1

TH #	Transients Including Valve Reclosure System Failure	Operator Action	HZP	IEF	Percent Contribution to Total Frequency of Crack Initiation (FCI)				Percent Contribution to Total Through Wall Cracking Frequency (TWCF)				Mean CPI				Mean CPTWC			
					32	60	Ext-A	Ext-B	32	60	Ext-A	Ext-B	32	60	Ext-A	Ext-B	32	60	Ext-A	Ext-B
122	Stuck-open pressurizer safety valve. Valve recloses at 6000 secs.	Operator throttles HPI at 10 minutes after 2.7 K [5°F] subcooling and 254-cm [100"] pressurizer level is reached (throttling criteria is 27.8 K [50°F] subcooling).	Y	7.57E-06	13.7	4.7	0.4	0.3	76.8	74.5	26.9	11.27	3.01E-06	7.23E-06	4.68E-05	1.44E-04	3.01E-06	7.23E-06	4.68E-05	1.44E-04
165	Stuck-open pressurizer safety valve. Valve recloses at 6000 secs [RCS low pressure point].	None	Y	1.76E-06	4.0	1.3	0.1	0.1	22.4	20.3	7.8	2.26	2.75E-06	6.55E-06	4.24E-05	1.24E-04	2.75E-06	6.55E-06	4.24E-05	1.24E-04
124	Stuck-open pressurizer safety valve. Valve recloses at 3000 secs.	Operator throttles HPI at 10 minutes after 2.7 K [5°F] subcooling and 254-cm [100"] pressurizer level is reached (throttling criteria is 27.8 K [50°F] subcooling).	Y	7.57E-06	0.1	0.2	0.2	0.2	0.5	3.0	12.8	9.02	3.61E-08	2.81E-07	1.48E-05	9.38E-05	3.60E-08	2.80E-07	1.48E-05	9.37E-05
168	TT/RT with stuck-open pzr SRV. SRV assumed to reclose at 3000 secs.	None	Y	1.76E-06	0.0	0.0	0.0	0.0	0.1	0.4	2.4	1.33	5.48E-08	3.75E-07	1.78E-05	1.10E-04	5.45E-08	3.73E-07	1.78E-05	1.09E-04
113	Stuck-open pressurizer safety valve. Valve recloses at 6000 secs.	After valve recloses, operator throttles HPI 10 minutes after 2.7 K [5°F] subcooling and 254-cm [100"] pressurizer level is reached (throttling criteria is 27.8 K [50°F] subcooling)	N	5.07E-05	0.0	0.0	0.0	0.0	0.0	0.0	0.0	0.30	0.00E+00	0.00E+00	0.00E+00	1.42E-07	0.00E+00	0.00E+00	0.00E+00	1.31E-07
109	Stuck-open pressurizer safety valve. Valve recloses at 6000 secs [RCS low pressure point].	None	N	9.58E-06	0.0	0.0	0.0	0.0	0.0	0.0	0.0	0.00	0.00E+00	0.00E+00	1.31E-09	1.83E-07	0.00E+00	0.00E+00	1.30E-09	1.83E-07
112	Stuck-open pressurizer safety valve. Valve recloses at 6000 secs.	After valve recloses, operator throttles HPI 1 minute after 2.7 K [5°F] subcooling and 254-cm [100"] pressurizer level is reached (throttling criteria is 27 K [50°F] subcooling)	N	1.25E-04	-	-	-	-	-	-	-	-	-	-	-	-	-	-	-	-
114	Stuck-open pressurizer safety valve. Valve recloses at 3000 secs.	After valve recloses, operator throttles HPI 1 minute after 2.7 K [5°F] subcooling and 254-cm [100"] pressurizer level is reached (throttling criteria is 50°F subcooling)	N	1.25E-04	-	-	-	-	-	-	-	-	-	-	-	-	-	-	-	-

TH #	Transients Including Valve Reclosure System Failure	Operator Action	HZP	IEF	Percent Contribution to Total Frequency of Crack Initiation (FCI)				Percent Contribution to Total Through Wall Cracking Frequency (TWCF)				Mean CPI				Mean CPTWC			
					32	60	Ext-A	Ext-B	32	60	Ext-A	Ext-B	32	60	Ext-A	Ext-B	32	60	Ext-A	Ext-B
115	Stuck-open pressurizer Safety Valve. Valve recloses at 3000 secs.	After valve recloses, operator throttles HPI 10 minutes after 2.7 K [5°F] subcooling and 254-cm [100"] pressurizer level is reached (throttling criteria is 50°F subcooling)	N	5.07E-05	-	-	-	-	-	-	-	-	-	-	-	-	-	-	-	-
121	Stuck-open pressurizer safety valve. Valve recloses at 6000 secs .	Operator throttles HPI at 1 minute after 2.7 K [5°F] subcooling and 254-cm [100"] pressurizer level is reached [throttling criteria is 27.8 K [50°F] subcooling].	Y	2.28E-05	0.0	0.0	0.0	0.0	0.0	0.0	0.0	0.00	0.00E+00	0.00E+00	6.54E-11	2.06E-07	0.00E+00	0.00E+00	0.00E+00	1.28E-08
123	Stuck-open pressurizer safety valve. Valve recloses at 3000 secs.	Operator throttles HPI at 1 minute after 2.7 K [5°F] subcooling and 254-cm [100"] pressurizer level is reached (throttling criteria is 27.8 K [50°F] subcooling).	Y	2.28E-05	0.0	0.0	0.0	0.0	0.0	0.0	0.0	0.00	0.00E+00	0.00E+00	6.54E-11	2.06E-07	0.00E+00	0.00E+00	0.00E+00	1.28E-08
149	TT/RT with stuck-open pzr SRV. SRV assumed to reclose at 3000 secs.	None	N	9.58E-06	-	-	-	-	-	-	-	-	-	-	-	-	-	-	-	-
Palisades																				
65	One stuck-open pressurizer SRV that recloses at 6000 sec after initiation. Containment spray is assumed not to actuate.	None	Y	1.24E-04	6.5	5.8	4.4	2.4	67.2	45.4	17.5	8.40	2.60E-05	5.50E-05	2.57E-04	8.40E-04	2.53E-05	5.40E-05	2.55E-04	8.37E-04
48	Two stuck-open pressurizer SRVs that reclose at 6000 sec after initiation. Containment spray is assumed not to actuate.	None	Y	7.67E-07	0.1	0.1	0.1	0.0	1.4	0.9	0.3	0.12	8.57E-05	1.67E-04	6.50E-04	1.96E-03	8.46E-05	1.66E-04	6.47E-04	1.95E-03
53	Turbine/reactor trip with two stuck-open pressurizer SRVs that reclose at 6000 sec after initiation. Containment spray is assumed not to actuate.	None	N	1.09E-03	0.0	0.0	0.3	0.4	0.0	0.2	0.8	1.27	9.91E-10	3.34E-08	1.48E-06	1.23E-05	3.86E-10	1.62E-08	1.13E-06	1.15E-05

TH #	Transients Including Valve Reclosure System Failure	Operator Action	HZP	IEF	Percent Contribution to Total Frequency of Crack Initiation (FCI)				Percent Contribution to Total Through Wall Cracking Frequency (TWCF)				Mean CPI				Mean CPTWC			
					32	60	Ext-A	Ext-B	32	60	Ext-A	Ext-B	32	60	Ext-A	Ext-B	32	60	Ext-A	Ext-B
42	Turbine/reactor trip with two stuck-open pressurizer SRVs. Containment spray is assumed not to actuate.	Operator assumed to throttle HPI if auxiliary feedwater is running with SG wide range level > -84% and RCS subcooling > 25 F. HPI is throttled to maintain pressurizer level between 40 and 60%.	N	7.67E-07	-	-	-	-	-	-	-	-	-	-	-	-	-	-	-	-

Table A.4. Transient descriptions and FAVOR 04.1 results for stuck-open primary valve transients (no value reclosure)

TH #	Transients Without Valve Reclosure System Failure	Operator Action	HZP	IEF	Percent Contribution to Total Frequency of Crack Initiation (FCI)				Percent Contribution to Total Through Wall Cracking Frequency (TWCF)				Mean CPI				Mean CPTWC			
					32	60	Ext-A	Ext-B	32	60	Ext-A	Ext-B	32	60	Ext-A	Ext-B	32	60	Ext-A	Ext-B
Beaver Valley Unit 1																				
14	Reactor/turbine trip w/one stuck-open pressurizer SRV	None.	N	2.23E-04	0.0	0.0	0.0	0.0	0.0	0.0	0.0	0.00	1.56E-11	2.02E-08	9.33E-08	2.93E-06	0.00E+00	0.00E+00	1.09E-15	3.80E-10
34	Reactor/turbine trip w/two stuck-open pressurizer SRVs	None.	N	4.95E-07	0.0	0.0	0.0	0.0	0.0	0.0	0.0	0.00	2.39E-07	2.53E-06	1.41E-05	1.53E-04	1.60E-17	1.64E-09	1.56E-08	1.03E-06
64	Reactor/turbine trip w/two stuck-open pressurizer SRVs	None.	Y	8.67E-08	0.0	0.0	0.0	0.0	0.0	0.0	0.0	0.00	7.51E-06	4.39E-05	2.18E-04	1.80E-03	1.28E-09	7.65E-08	8.92E-07	2.43E-05
94	Reactor/turbine trip w/one stuck-open pressurizer SRV.	None.	Y	4.10E-05	0.0	0.0	0.0	0.0	0.0	0.0	0.0	0.00	0.00E+00	4.70E-11	2.14E-08	7.76E-06	0.00E+00	0.00E+00	0.00E+00	4.73E-08
Oconee Unit 1																				
169	TT/RT with stuck-open pzr SRV [valve flow area reduced by 30%]. Summer conditions assumed [HPI, LPI temp = 302 K [85° F] and CFT temp = 310 K [100° F]]. Vent valves do not function.	None	Y	7.33E-06	0.0	0.0	0.1	0.2	0.0	0.0	0.6	0.95	1.90E-10	2.65E-08	1.47E-05	9.92E-05	0.00E+00	3.12E-11	7.13E-07	1.00E-05
146	TT/RT with stuck-open pzr SRV [valve flow area reduced by 30%]. Summer conditions assumed [HPI, LPI temp = 302 K [85° F] and CFT temp = 310 K [100° F]]. Vent valves do not function.	None	N	4.23E-05	0.0	0.0	0.0	0.0	0.0	0.0	0.0	0.17	0.00E+00	0.00E+00	9.28E-08	3.91E-06	0.00E+00	0.00E+00	7.44E-09	8.90E-07
147	TT/RT with stuck-open pzr SRV. Summer conditions assumed [HPI, LPI temp = 302 K [85° F] and CFT temp = 310 K [100° F]].	None	N	3.63E-05	-	-	-	-	-	-	-	-	-	-	-	-	-	-	-	-

TH #	Transients Without Valve Reclosure System Failure	Operator Action	HZP	IEF	Percent Contribution to Total Frequency of Crack Initiation (FCI)				Percent Contribution to Total Through Wall Cracking Frequency (TWCF)				Mean CPI				Mean CPTWC			
					32	60	Ext-A	Ext-B	32	60	Ext-A	Ext-B	32	60	Ext-A	Ext-B	32	60	Ext-A	Ext-B
148	TT/RT with partially stuck-open pzr SRV [flow area equivalent to 1.5 in diameter opening]. HTC coefficients increased by 1.3.	None	N	4.23E-05	-	-	-	-	-	-	-	-	-	-	-	-	-	-	-	-
170	TT/RT with stuck-open pzr SRV. Summer conditions assumed [HPI, LPI temp = 302 K [85° F] and CFT temp = 310 K [100° F]].	None	Y	6.28E-06	-	-	0.0	0.0	-	-	-	0.00	-	-	6.68E-12	1.38E-07	-	-	-	7.72E-09
171	TT/RT with partially stuck-open pzr SRV [flow area equivalent to 1.5 in diameter opening]. HTC coefficients increased by 1.3.	None	Y	7.33E-06	-	-	-	-	-	-	-	-	-	-	-	-	-	-	-	-

Table A.5. Transient descriptions and FAVOR 04.1 results for large-diameter steam line break transients

TH Transient #	System Failure	Operator Action	HZP	IEF	Percent Contribution to Total Frequency of Crack Initiation (FCI)				Percent Contribution to Total Through Wall Cracking Frequency (TWCF)				Mean CPI				Mean CPTWC			
					32	60	Ext-A	Ext-B	32	60	Ext-A	Ext-B	32	60	Ext-A	Ext-B	32	60	Ext-A	Ext-B
Beaver Valley Unit 1																				
103	Main steam line break with AFW continuing to feed affected generator for 30 minutes.	Operator controls HHSI 30 minutes after allowed. Break is assumed to occur inside containment so that the operator trips the RCPs due to adverse containment conditions.	Y	1.1E-05	0.07	0.17	0.30	0.63	0.74	2.54	4.78	5.17	7.36E-06	6.67E-05	4.14E-04	4.68E-03	3.96E-07	8.57E-06	7.41E-05	1.29E-03
104	Main steam line break with AFW continuing to feed affected generator for 30 minutes.	Operator controls HHSI 60 minutes after allowed. Break is assumed to occur inside containment so that the operator trips the RCPs due to adverse containment conditions.	N	1.1E-04	0.07	0.25	0.36	1.72	0.01	0.55	2.67	10.24	3.63E-07	6.82E-06	5.76E-05	1.22E-03	5.21E-10	1.47E-07	4.93E-06	2.36E-04
102	Main steam line break with AFW continuing to feed affected generator for 30 minutes.	Operator controls HHSI 30 minutes after allowed. Break is assumed to occur inside containment so that the operator trips the RCPs due to adverse containment conditions.	N	1.0E-04	0.03	0.11	0.36	1.62	0.00	0.32	3.05		3.63E-07	6.82E-06	5.76E-05	1.22E-03	5.21E-10	1.47E-07	4.93E-06	2.36E-04
107	Main steam line break with AFW continuing to feed affected generator.	Operator controls HHSI 30 minutes after allowed. Break is assumed to occur inside containment so that the operator trips the RCPs due to adverse containment conditions.	Y	4.3E-07	0.00	0.00	0.01	0.02	0.03	0.12	0.18	0.22	6.16E-06	5.73E-05	3.48E-04	3.95E-03	5.28E-07	1.03E-05	8.51E-05	1.35E-03
105	Main steam line break with AFW continuing to feed affected generator for 30 minutes.	Operator controls HHSI 60 minutes after allowed. Break is assumed to occur inside containment so that the operator trips the RCPs due to adverse containment conditions.	Y	1.1E-05	0	0	0.01	0.02	0	0.01	0.04	0.14	8.62E-09	6.03E-07	5.76E-06	1.45E-03	2.70E-11	3.75E-08	8.38E-07	363E-05
106	Main steam line break with AFW continuing to feed affected generator.	Operator controls HHSI 30 minutes after allowed. Break is assumed to occur inside containment so that the operator trips the RCPs due to adverse containment conditions.	N	2.2E-06	0	0	0.01	0.04	0	0	0.08	0.21	3.52E-07	6.92E-06	5.83E-05	1.23E-03	2.79E-10	1.33E-07	4.73E-06	2.35E-04
74	Main steam line break with AFW continuing to feed affected generator	None.	N	1.5E-06	0	0	0	0.01	0	0	0	0.02	1.46E-08	7.29E-07	5.94E-06	1.47E-04	4.12E-12	3.20E-08	8.04E-07	3.66E-05

TH Transient #	System Failure	Operator Action	HZP	IEF	Percent Contribution to Total Frequency of Crack Initiation (FCI)				Percent Contribution to Total Through Wall Cracking Frequency (TWCF)				Mean CPI				Mean CPTWC			
					32	60	Ext-A	Ext-B	32	60	Ext-A	Ext-B	32	60	Ext-A	Ext-B	32	60	Ext-A	Ext-B
81	Main Steam Line Break with AFW continuing to feed affected generator and with HHSI failure initially.	Operator opens ADVs (on intact generators). HHSI is restored after CFTs discharge 50%.	N	2.7E-06	0	0	0	0	0	0	0	0	0.00E+00	3.56E-13	3.54E-09	2.75E-06	0.00E+00	0.00E+00	0.00E+00	1.24E-09
Oconee Unit 1																				
27	MSLB without trip of turbine-driven emergency feedwater.	Operator throttles HPI to maintain 27.8 K [50° F] subcooling margin.	N	2.1E-06	0	0	0	0	0	0	0	0.01	0.00E+00	3.05E-13	2.80E-07	5.11E-06	0.00E+00	0.00E+00	5.16E-09	3.70E-07
99	MSLB with trip of turbine-driven EFW by MSLB Circuitry	HPI is throttled 20 minutes after 2.7 K [5°F] subcooling and 254-cm [100"] pressurizer level is reached (throttling criteria is 27.8 K [50°F] subcooling).	N	2.4E-07	0	0	0	0	0	0	0	0	0.00E+00	0.00E+00	1.19E-07	2.08E-06	0.00E+00	0.00E+00	2.06E-09	4.26E-07
100	MSLB with trip of turbine-driven EFW by MSLB Circuitry	Operator throttles HPI 20 minutes after 2.7 K [5°F] subcooling and 254-cm [100"] pressurizer level is reached (throttling criteria is 27.8 K [50°F] subcooling).	Y	5.1E-08	0	0	0	0	0	0	0	0	0.00E+00	0.00E+00	7.81E-08	6.06E-06	0.00E+00	0.00E+00	4.08E-08	3.47E-06
101	MSLB without trip of turbine-driven EFW by MSLB Circuitry	Operator throttles HPI to maintain 27.8 K [50° F] subcooling margin (throttling criteria is 27.8 K [50°F] subcooling).	Y	3.9E-07	0	0	0	0	0	0	0	0	0.00E+00	1.51E-09	1.05E-06	7.58E-06	0.00E+00	0.00E+00	1.06E-09	2.02E-07
Palisades																				
54	Main steam line break with failure of both MSIVs to close. Break assumed to be inside containment causing containment spray actuation.	Operator does not isolate AFW on affected SG. Operator does not throttle HPI.	N	4.3E-06	0.44	0.51	0.66	0.69	1.66	1.88	1.62	1.51	5.16E-05	1.37E-04	9.41E-04	4.88E-03	1.79E-05	6.30E-05	6.12E-04	3.52E-03
27	Main steam line break with controller failure resulting in the flow from two AFW pumps into affected steam generator. Break assumed to be inside containment causing containment spray actuation.	Operator starts second AFW pump.	N	3.7E-05	0.13	0.2	0.37	0.41	0.26	0.41	0.72	0.92	1.56E-06	5.92E-06	7.00E-05	4.71E-04	2.97E-07	1.48E-06	3.23E-05	2.98E-04

26	Main steam line break with the break assumed to be inside containment causing containment spray actuation.	Operator isolates AFW to affected SG at 30 minutes after initiation.	N	5.7E-04	0.04	0.11	0.41	0.7	0.02	0.05	0.29	0.98	4.98E-08	2.41E-07	5.20E-06	5.02E-05	2.13E-09	1.55E-08	1.01E-06	2.07E-05
50	Main steam line break with controller failure resulting in the flow from two AFW pumps into affected steam generator. Break assumed to be inside containment causing containment spray actuation.	Operator starts second AFW pump. Operator does not throttle HPI.	Y	5.8E-07	0.01	0.01	0.01	0.01	0.02	0.02	0.03	0.03	4.45E-06	1.45E-05	1.41E-04	8.92E-04	1.24E-06	4.73E-06	7.46E-05	6.03E-04
51	Main steam line break with failure of both MSIVs to close. Break assumed to be inside containment causing containment spray actuation.	Operator does not isolate AFW on affected SG. Operator does not throttle HPI.	Y	7.5E-08	0	0	0.01	0.01	0.02	0.02	0.02	0.02	2.55E-05	7.09E-05	5.26E-04	2.85E-03	9.92E-06	3.51E-05	3.53E-04	2.05E-03
34	Main steam line break concurrent with a single tube failure in SG-A due to MSLB vibration.	Operator isolates AFW to affected SG at 15 minutes after initiation. Operator trips RCPs assuming that they do not trip as a result of the event. Operator assumed to throttle HPI if auxiliary feedwater is running with SG wide range level > -84% and RCS	N	1.5E-05	0	0.01	0.03	0.04	0	0.01	0.03	0.06	1.43E-07	7.88E-07	1.38E-05	1.08E-04	8.33E-09	6.97E-08	3.30E-06	4.77E-05
49	Main steam line break with the break assumed to be inside containment causing containment spray actuation.	Operator isolates AFW to affected SG at 30 minutes after initiation. Operator does not throttle HPI.	Y	1.0E-05	0	0	0	0.01	0	0	0	0.01	9.02E-09	5.39E-08	1.83E-06	1.98E-05	5.90E-10	6.17E-09	3.85E-07	9.10E-06
24	Main steam line break with the break assumed to be inside containment causing containment spray actuation.	None	N	2.4E-06	0	0	0	0	0	0	0	0	4.98E-08	2.41E-07	5.20E-06	5.02E-05	1.67E-09	1.41E-08	9.22E-07	1.96E-05
29	Main steam line break with break assumed to be inside containment causing containment spray actuation.	None. Operator does not throttle HPI.	Y	4.2E-08	0	0	0	0	0	0	0	0	2.02E-07	5.18E-07	3.54E-06	2.36E-05	1.94E-07	4.71E-07	2.16E-06	1.38E-05

Table A.6. Transient descriptions and FAVOR 04.1 results for SO-2 transients involving all (or a very large number of) stuck-open valves

TH Transient #	System Failure	Operator Action	HZP	IEF	Percent Contribution to Total Frequency of Crack Initiation (FCI)				Percent Contribution to Total Through Wall Cracking Frequency (TWCF)				Mean CPI				Mean CPTWC			
					32	60	Ext-A	Ext-B	32	60	Ext-A	Ext-B	32	60	Ext-A	Ext-B	32	60	Ext-A	Ext-B
Beaver Valley Unit 1																				
110	Small steam line break (simulated by sticking open all SG-A SRVs) with AFW continuing to feed affected generator for 30 minutes	Operator controls HHSI 60 minutes after allowed.	N	6.9E-04	0	0	0.02	0.32	0	0	0.02	1.43	1.75E-11	2.48E-08	4.71E-07	3.64E-05	0.00E+00	0.00E+00	4.20E-09	3.65E-06
108	Small steam line break (simulated by sticking open all SG-A SRVs) with AFW continuing to feed affected generator for 30 minutes.	Operator controls HHSI 30 minutes after allowed.	Y	6.5E-04	0	0.01	0.01	0.2	0	0	0.02	0.64	1.75E-11	2.48E-08	4.71E-07	3.64E-05	0.00E+00	0.00E+00	4.20E-09	3.65E-06
111	Small steam line break (simulated by sticking open all SG-A SRVs) with AFW continuing to feed affected generator for 30 minutes.	Operator controls HHSI 60 minutes after allowed. Break is assumed to occur inside containment so that the operator trips the RCPs due to adverse containment conditions.	Y	6.8E-05	0	0.01	0.02	0.11	0	0	0.07	0.51	3.08E-09	2.62E-07	3.45E-06	1.40E-04	1.06E-15	5.14E-10	1.43E-07	2.04E-05
109	Small steam line break (simulated by sticking open all SG-A SRVs) with AFW continuing to feed affected generator for 30 minutes.	Operator controls HHSI 30 minutes after allowed. Break is assumed to occur inside containment so that the operator trips the RCPs due to adverse containment conditions.	Y	6.8E-05	0	0	0.02	0.12	0	0	0.03	0.43	3.08E-09	2.62E-07	3.45E-06	1.40E-04	1.06E-15	5.14E-10	1.43E-07	2.04E-05
112	Small steam line break (simulated by sticking open all SG-A SRVs) with AFW continuing to feed affected generator.	Operator controls HHSI 30 minutes after allowed. Break is assumed to occur inside containment so that the operator trips the RCPs due to adverse containment conditions.	N	1.4E-05	0	0	0	0.01	0	0	0	0.02	1.86E-11	2.13E-08	3.84E-07	3.55E-05	0.00E+00	3.35E-12	4.82E-09	3.76E-06
118	Small steam line break (simulated by sticking open all SG-A SRVs) with AFW continuing to feed affected generator	None.	N	9.3E-06	0	0	0	0.01	0	0	0	0.02	3.14E-10	5.80E-08	1.02E-06	5.40E-05	0.00E+00	2.34E-11	2.44E-08	6.94E-06

TH Transient #	System Failure	Operator Action	HZP	IEF	Percent Contribution to Total Frequency of Crack Initiation (FCI)				Percent Contribution to Total Through Wall Cracking Frequency (TWCF)				Mean CPI				Mean CPTWC			
					32	60	Ext-A	Ext-B	32	60	Ext-A	Ext-B	32	60	Ext-A	Ext-B	32	60	Ext-A	Ext-B
113	Small steam line break (simulated by sticking open all SG-A SRVs) with AFW continuing to feed affected generator.	Operator controls HHSI 30 minutes after allowed. Break is assumed to occur inside containment so that the operator trips the RCPs due to adverse containment conditions.	Y	2.7E-06	0	0	0	0	0	0	0	0.01	4.07E-10	2.35E-07	3.06E-06	1.00E-04	7.76E-14	6.47E-10	4.69E-08	1.17E-05
78	Reactor/turbine trip with failure of MFW and AFW.	Operator opens all ASDVs to let condensate fill SGs.	N	3.3E-8	0	0	0	0	0	0	0	0.00	0	0	0	0	0	0	0	1.0E-19
Oconee Unit 1																				
89	Reactor/turbine trip with loss of MFW and EFW.	Operator opens all TBVs to depressurize the secondary side to below the condensate booster pump shutoff head so that these pumps feed the steam generators. Booster pumps are assumed to be initially uncontrolled so that the steam generators are overfilled	N	5.4E-07	0	0	0	0	0	0	0	0	0.00E+00	0.00E+00	0.00E+00	0.00E+00	0.00E+00	0.00E+00	0.00E+00	0.00E+00
98	Reactor/turbine trip with loss of MFW and EFW	Operator opens all TBVs to depressurize the secondary side to below the condensate booster pump shutoff head so that these pumps feed the steam generators. Booster pumps are assumed to be initially uncontrolled so that the steam generators are overfilled	Y	1.0E-07	0	0	0	0	0	0	0	0	0.00E+00	0.00E+00	0.00E+00	0.00E+00	0.00E+00	0.00E+00	0.00E+00	0.00E+00
Palisades																				
22	Turbine/reactor trip with loss of MFW and AFW.	Operator depressurizes through ADVs and feeds SGs using condensate booster pumps. Operators maintain a cooldown rate within technical specification limits and throttle condensate flow at 84% level in the steam generator.	N	6.7E-05	0	0	0	0	0	0	0	0.01	0.00E+00	1.38E-12	3.12E-08	1.20E-06	0.00E+00	4.25E-13	5.61E-09	7.07E-07

Table A.7. Transient descriptions and FAVOR 04.1 results for SO-2 transients involving just a few (one or two) stuck-open secondary valves

TH Transient #	System Failure	Operator Action	HZP	IEF	Percent Contribution to Total Frequency of Crack Initiation (FCI)				Percent Contribution to Total Through Wall Cracking Frequency (TWCF)				Mean CPI				Mean CPTWC			
					32	60	Ext-A	Ext-B	32	60	Ext-A	Ext-B	32	60	Ext-A	Ext-B	32	60	Ext-A	Ext-B
Beaver Valley Unit 1																				
No transients of this type were analyzed																				
Oconee Unit 1																				
28	Reactor/turbine trip with 1 stuck-open safety valve in SG-A.	None	N	7.5E-08	0	0	0	0	0	0	0	0	0	0	0	0	0	0	0	0
29	Reactor/turbine trip with 1 stuck-open safety valve in SG-A and a second stuck-open safety valve in SG-B.	None	N	3.1E-07	0	0	0	0	0	0	0	0	0	0	0	0	0	0	0	0
30	Reactor/turbine trip with 1 stuck-open safety valve in SG-A.	None	Y	1.5E-07	0	0	0	0	0	0	0	0	0	0	0	0	0	0	0	0
31	Reactor/turbine trip with 1 stuck-open safety valve in SG-A and a second stuck-open safety valve in SG-B.	None	Y	8.4E-09	0	0	0	0	0	0	0	0	0	0	0	6.2E-11	0	0	0	6.2E-11
36	Reactor/turbine trip with 1 stuck-open safety valve in SG-A and a second stuck-open safety valve in SG-B.	Operator throttles HPI to maintain 27.8 K [50° F] subcooling and 304.8-cm [120-in.] pressurizer level.	N	1.4E-05	0	0	0	0	0	0	0	0	0	0	0	0	0	0	0	0
37	Reactor/turbine trip with 1 stuck-open safety valve in SG-A.	Operator throttles HPI to maintain 27.8 K [50° F] subcooling and 304.8-cm [120-in.] pressurizer level.	Y	1.4E-06	0	0	0	0	0	0	0	0	0	0	0	0	0	0	0	0
38	Reactor/turbine trip with 1 stuck-open safety valve in SG-A and a second stuck-open safety valve in SG-B.	Operator throttles HPI to maintain 27.8 K [50° F] subcooling and 304.8-cm [120-in.] pressurizer level.	Y	2.7E-06	0	0	0	0	0	0	0	0	0	0	0	0	0	0	0	0

TH Transient #	System Failure	Operator Action	HZP	IEF	Percent Contribution to Total Frequency of Crack Initiation (FCI)				Percent Contribution to Total Through Wall Cracking Frequency (TWCF)				Mean CPI				Mean CPTWC			
					32	60	Ext-A	Ext-B	32	60	Ext-A	Ext-B	32	60	Ext-A	Ext-B	32	60	Ext-A	Ext-B
90	Reactor/turbine trip with 2 stuck-open safety valves in SG-A.	Operator throttles HPI 20 minutes after 2.7 K [5°F] subcooling and 254-cm [100"] pressurizer level is reached [throttling criteria is 27.8 K [50°F] subcooling].	N	6.3E-07	0	0	0	0	0	0	0	0	0	0	0	0	0	0	0	0
102	Reactor/turbine trip with 2 stuck-open safety valves in SG-A.	Operator throttles HPI 20 minutes after 2.77 K [5°F] subcooling and 254-cm [100-in.] pressurizer level is reached (throttling criteria is 27 K [50°F] subcooling).	Y	2.0E-07	0	0	0	0	0	0	0	0	0	0	0	0	0	0	0	0
Palisades																				
P55	Turbine/reactor trip with 2 stuck-open ADVs on SG-A combined with controller failure resulting in the flow from two AFW pumps into affected steam generator.	Operator starts second AFW pump.	N	2.7E-03	0.6	1.7	5.0	5.4	3.0	8.4	16.8	18.0	3.5E-08	3.0E-07	5.9E-06	3.8E-05	1.9E-08	1.9E-07	5.0E-06	3.6E-05
19	Reactor trip with 1 stuck-open ADV on SG-A.	None. Operator does not throttle HPI.	Y	2.3E-03	0.2	0.8	2.5	3.1	0.9	3.2	7.4	9.7	9.1E-09	1.1E-07	2.4E-06	1.6E-05	3.6E-09	5.4E-08	1.9E-06	1.5E-05
52	Reactor trip with 1 stuck-open ADV on SG-A. Failure of both MSIVs (SG-A and SG-B) to close.	Operator does not isolate AFW on affected SG. Normal AFW flow assumed (200 gpm). Operator does not throttle HPI.	Y	6.4E-04	0	0.0	0.1	0.4	0.0	0.1	0.4	1.2	1.7E-08	1.5E-07	3.0E-06	1.9E-05	6.6E-09	7.6E-08	2.3E-06	1.8E-05
16	Turbine/reactor trip with 2 stuck-open ADVs on SG-A combined with controller failure resulting in the flow from two AFW pumps into affected steam generator.	Operator starts second AFW pump. Operator isolates AFW to affected SG at 30 minutes after initiation. Operator assumed to throttle HPI if auxiliary feedwater is running with SG wide range level > -84% and RCS subcooling > 25 F. HPI is throttled to main	N	1.2E-04	0	0	0	0	0	0	0	0	0	9.4E-12	2.3E-08	8.6E-07	0	6.4E-13	2.1E-09	2.7E-07

TH Transient #	System Failure	Operator Action	HZP	IEF	Percent Contribution to Total Frequency of Crack Initiation (FCI)				Percent Contribution to Total Through Wall Cracking Frequency (TWCF)				Mean CPI				Mean CPTWC			
					32	60	Ext-A	Ext-B	32	60	Ext-A	Ext-B	32	60	Ext-A	Ext-B	32	60	Ext-A	Ext-B
18	Turbine/reactor trip with 1 stuck-open ADV on SG-A. Failure of both MSIVs (SG-A and SG-B) to close.	Operator does not isolate AFW on affected SG. Normal AFW flow assumed (200 gpm). Operator assumed to throttle HPI if auxiliary feedwater is running with SG wide range level > -84% and RCS subcooling > 25 F. HPI is throttled to maintain pressurizer level	N	4.7E-03	0	0	0	0	0	0	0	0	0	2.1E-12	3.0E-11	8.3E-09	0	8.8E-13	2.9E-11	3.3E-09

Table A.8. Transient descriptions and FAVOR 04.1 results for feed and bleed, overfeed, and steam generator tube rupture transients

Plant	Class	TH #	System Failure	Operator Action	HZP	IEF	Percent Contribution to Total Frequency of Crack Initiation (FCI)				Percent Contribution to Total Through Wall Cracking Frequency (TWCF)				Mean CPI				Mean CPTWC			
							32	60	Ext-A	Ext-B	32	60	Ext-A	Ext-B	32	60	Ext-A	1000	32	60	Ext-A	Ext-B
Palisades	F&B	31	Turbine/reactor trip with failure of MFW and AFW. Containment spray actuation assumed due to PORV discharge.	Operator maintains core cooling by "feed and bleed" using HPI to feed and two PORVs to bleed.	N	1.29E-05	0.05	0.05	0.07	0.07	0.06	0.08	0.10	0.09	2.09E-06	5.34E-06	3.82E-05	2.04E-04	2.36E-07	1.08E-06	1.54E-05	9.24E-05
Palisades	F&B	32	Turbine/reactor trip with failure of MFW and AFW. Containment spray actuation assumed due to PORV discharge.	Operator maintains core cooling by "feed and bleed" using HPI to feed and two PORV to bleed. AFW is recovered 15 minutes after initiation of "feed and bleed" cooling. Operator closes PORVs when SG level reaches 60%.	N	1.08E-06	0.00	0.00	0.00	0.00	0.00	0.00	0.00	0.00	3.87E-09	6.77E-08	2.26E-06	1.70E-05	1.50E-09	3.30E-08	1.71E-06	1.58E-05
Beaver	F&B	31	Reactor/turbine trip w/ feed & bleed	Operator opens all pzr PORVs & uses all charging/HHSI pumps	N	3.10E-7	0.00	0.00	0.00	0.00	0.00	0.00	0.00	0.00	3.44E-7	3.28E-6	1.76E-5	1.72E-4	3.10E-12	4.02E-9	3.32E-8	1.39E-6
Beaver	Overfeed	76	Reactor/turbine trip w/full MFW to all 3 SGs (MFW maintains SG level near top).	Operator trips reactor coolant pumps.	Y	1.1E-04	0	0	0.00	0.00	0	0	0.00	0.00	0	0	1.14E-10	1.74E-6	0	0	8.92E-15	9.73E-08
Oconee	SGTR	127	SGTR with a stuck-open SRV in SG-B. A reactor trip is assumed to occur at the time of the tube rupture. Stuck safety relief valve is assumed to reclose 10 minutes after initiation.	Operator trips RCPs 1 minute after initiation. Operator also throttles HPI 10 minutes after 2.77 K [5° F] subcooling and 254-cm [100-in.] pressurizer level is reached (assumed throttling criteria is 27 K [50°F] subcooling).	Y	1.25E-07	-	-	-	-	-	-	-	-	-	-	-	-	-	-	-	-

Table A.9. Transient descriptions and FAVOR 04.1 results for mixed primary and secondary initiator transients

#	System Failure	Operator Action	HZP	IEF	Percent Contribution to Total Frequency of Crack Initiation (FCI)				Percent Contribution to Total Through Wall Cracking Frequency (TWCF)				Mean CPI				Mean CPTWC			
					32	60	Ext-A	Ext-B	32	60	Ext-A	Ext-B	32	60	Ext-A	Ext-B	32	60	Ext-A	Ext-B
Beaver Valley Unit 1																				
B-65	Reactor/turbine trip w/two stuck-open pressurizer SRVs and HHSI failure	Operator opens all ASDVs 5 minutes after HHSI would have come on.	N	1.04E-09	0.0	0.0	0.0	0.0	0.0	0.0	0.0	0.00	1.71E-09	1.43E-07	1.19E-06	3.64E-05	0.00E+00	0.00E+00	6.42E-14	2.98E-08
68	Reactor/turbine trip w/two stuck-open pressurizer SRVs that reclose at 6000 s with HHSI failure.	Operator opens all ASDVs 5 minutes after HHSI would have come on.	N	1.33E-08	0.0	0.0	0.0	0.0	0.0	0.0	0.0	0.00	1.41E-07	5.69E-07	2.36E-06	5.94E-05	1.36E-07	2.69E-07	1.16E-07	1.51E-06
72	Reactor/turbine trip w/one stuck-open pressurizer SRV with HHSI failure.	Operator opens all ASDVs 5 minutes after HHSI would have come on.	N	5.14E-07	-	-	-	-	-	-	-	-	-	-	-	1.19E-12	-	-	-	-
73	Reactor/turbine trip w/one stuck-open pressurizer SRV with HHSI failure	Operator open all ASDVs 5 minutes after HHSI would have come on.	Y	6.55E-08	-	-	-	-	-	-	-	-	-	-	-	4.64E-08	-	-	-	-
82	Reactor/turbine trip w/one stuck-open pressurizer SRV (recloses at 6000 s) and with HHSI failure.	Operator opens all ASDVs 5 minutes after HHSI would have started.	N	1.51E-06	0.0	0.0	0.0	0.0	0.0	0.0	0.0	0.00	0.00E+00	2.03E-11	0.00E+00	2.40E-09	0.00E+00	1.92E-11	0.00E+00	2.30E-09
83	2.54-cm [1.0-in.] surge line break with HHSI failure and motor driven AFW failure. MFW is tripped. Level control failure causes all steam generators to be overfed with turbine AFW, with the level maintained at top of SGs.	Operator trips RCPs. Operator opens all ASDVs 5 minutes after HHSI would have come on.	N	3.51E-06	-	-	-	-	-	-	-	-	-	-	-	-	-	-	-	-

#	System Failure	Operator Action	HZP	IEF	Percent Contribution to Total Frequency of Crack Initiation (FCI)				Percent Contribution to Total Through Wall Cracking Frequency (TWCF)				Mean CPI				Mean CPTWC			
					32	60	Ext-A	Ext-B	32	60	Ext-A	Ext-B	32	60	Ext-A	Ext-B	32	60	Ext-A	Ext-B
Oconee Unit 1																				
O-110	5.08-cm [2 in.] surge line break with HPI failure	At 15 minutes after transient initiation, operator opens both TBV to lower primary system pressure and allow CFT and LPI injection.	N	3.42E-06	0.18	0.46	1.18	1.48	0.00	0.00	0.64	1.16	7.19E-08	1.02E-06	2.62E-04	1.73E-03	4.93E-13	6.88E-10	2.16E-06	3.11E-05
120	2.54-cm [1-in.] surge line break with HPI Failure	At 15 minutes after sequence initiation, operators open all TBVs to depressurize the system to the CFT setpoint. When the CFTs are 50% discharged, HPI is assumed to be recovered. The TBVs are assumed remain opened for the duration of the transient	Y	4.22E-08	0.00	0.00	0.00	0.00	0.00	0.00	0.01	0.01	1.69E-12	7.80E-09	2.69E-06	2.50E-05	1.31E-12	7.03E-09	2.51E-06	2.28E-05
44	2.54-cm [1-in.] surge line break with HPI Failure	At 15 minutes after initiation, operators open all TBVs to depressurize the system to the CFT setpoint. When the CFTs are 50% discharged, HPI is assumed to be recovered. The TBVs are assumed remain open for the duration of the transient.	N	2.69E-07	0.00	0.00	0.00	0.00	0.00	-	-	0.01	0.00E+00	0.00E+00	4.41E-07	6.48E-06	-	-	4.27E-07	6.13E-06
119	2.54-cm [1-in.] surge line break with HPI Failure	At 15 minutes after transient initiation, the operator opens all turbine bypass valves to lower primary system pressure and allow core flood tank and LPI injection.	Y	4.41E-07	0.00	0.00	0.00	0.01	0.00	-	0.00	0.01	2.83E-10	1.36E-08	8.48E-06	5.52E-05	-	4.42E-14	1.81E-07	2.95E-06
8	2.54-cm [1-in.] surge line break with 1 stuck-open safety valve in SG-A.	None	N	9.68E-08	-	-	-	-	-	-	-	-	-	-	-	-	-	-	-	-
12	2.54-cm [1-in.] surge line break with 1 stuck-open safety valve in SG-A.	HPI throttled to maintain 27.8 K [50° F] subcooling margin	N	9.24E-07	-	-	-	-	-	-	-	-	-	-	-	-	-	-	-	-

15	2.54-cm [1-in.] surge line break with HPI Failure	At 15 minutes after transient initiation, operator opens all TBVs to lower primary system pressure and allow CFT and LPI injection.	N	3.39E-08	-	-	0.00	0.00	-	-	-	0.00	-	-	1.74E-09	5.79E-07	-	-	-	1.33E-09
111	2.54-cm [1-in.] surge line break with HPI failure	At 15 minutes after initiation, operator opens all TBVs to lower primary pressure and allow CFT and LPI injection. When the CFTs are 50% discharged, HPI is recovered. At 3000 seconds after initiation, operator starts throttling HPI to 55 K [100°F] subcooling	N	4.16E-07	-	-	0.00	0.00	-	-	-	0.00	-	-	1.01E-10	1.83E-07	-	-	-	2.07E-09
117	Stuck-open pressurizer safety valve and HPI failure	At 15 minutes after initiation, operator opens all TBV to lower primary pressure and allow CFT and LPI injection. When the CFTs are 50% discharged, HPI is recovered. The SRV is closed 5 minutes after HPI recovered. HPI is throttled at 1 minute after 2.	N	5.38E-07	0.0	0.0	0.0	0.0	0.0	0.0	0.0	0.01	2.12E-11	7.19E-09	4.21E-06	7.37E-05	0.00E+00	0.00E+00	8.73E-09	7.36E-07
116	Stuck-open pressurizer safety valve and HPI failure	At 15 minutes after initiation, operator opens all TBVs to lower primary pressure and allow CFT and LPI injection. When the CFTs are 50% discharged, HPI is recovered. The HPI is throttled 20 minutes after 2.7 K [5°F] subcooling and 254-cm [100"] pressure	N	2.60E-07	0.0	0.0	0.0	0.0	0.0	0.0	0.0	0.00	0.00E+00	0.00E+00	1.40E-07	6.43E-06	0.00E+00	0.00E+00	1.15E-10	6.68E-08
125	Stuck-open pressurizer safety valve and HPI Failure	At 15 minutes after initiation, operator opens all TBVs to lower primary pressure and allow CFT and LPI injection. When the CFTs are 50% discharged, HPI is recovered. HPI is throttled 20 minutes after 2.7 K [5°F] subcooling and 254-cm [100"] pressurize	Y	4.61E-08	0.0	0.0	0.0	0.0	0.0	0.0	0.0	0.00	1.44E-10	4.89E-09	6.24E-06	5.54E-05	0.00E+00	0.00E+00	7.42E-08	2.34E-06
126	Stuck-open pressurizer safety valve and HPI Failure	At 15 minutes after initiation, operator opens all TBVs to lower primary pressure and allow CFT and LPI injection. When the CFTs are 50% discharged, HPI is recovered. SRV is closed at 5 minutes after HPI is recovered. HPI is throttled at 1 minute after	Y	8.41E-08	0.0	0.0	0.0	0.0	0.0	0.0	0.0	0.00	0.00E+00	0.00E+00	5.26E-08	2.31E-06	0.00E+00	0.00E+00	7.67E-11	8.95E-08

Appendix B – Peer Review

Background

In response to a letter [Ref. 1] from the Chairman of the Advisory Committee on Reactor Safeguards (ACRS), the Executive Director for Operations (EDO) at the U.S. Nuclear Regulatory Commission (NRC) identified [Ref. 2] a need for conducting formal peer review of the developed technical basis for potential revision to PTS screening criteria in the PTS Rule (10 CFR 50.61).

Based on this mandate from the EDO, the Office of Nuclear Regulatory Research (RES) developed a scope of work and solicited a panel of experts to perform independent review of the developed technical basis. Peer review was carried out based on this work scope. Peer reviewers were selected for their expertise in each of the three key subject areas, namely probabilistic risk assessment (PRA) and human reliability analysis (HRA), thermal-hydraulics (TH) analysis and experimental validation, and probabilistic fracture mechanics (PFM) analysis and experimental validation.

Each peer reviewer was asked to provide their individual comments on the entire PTS technical basis without developing a consensus on a unified set of comments, so as to satisfy the requirements that this peer review panel is not a Federal Advisory Committee.

The following paragraphs address the objective and scope of peer review, peer review panel members and their fields of expertise, references, peer review comments, and RES responses to the peer review comments.

Objective

The objective of the peer review was to perform a review to assess the adequacy and reasonableness of the developed technical basis (as detailed by the draft PTS NUREG report and other supporting documents) to support a potential revision of the PTS Rule (10 CFR 50.61).

Scope

The scope of the peer review involved the following:

- Review the developed methodology, technical approach, data and results in the technical basis.
- Provide comments on the adequacy and reasonableness of the methodology used and the results obtained. This will involve assessing that the developed models, data and concepts are sufficient for their intended use. While we are not expecting the review panel to run the developed computer codes, the review panel may at their discretion perform calculations to check the validity of the results. Before undertaking such computations, any additional resources and level of effort (beyond what is authorized here) have to be requested in the form of a revision to the statement of work.
- The review process will compare the major elements of the PTS methodology against the desired characteristics and attributes that are elaborated in a PRA standard (e.g., ASME PRA Standard RA-S-2002 [Ref. 3]). The peer review will identify both strengths and weaknesses in the PTS

methodology. Key assumptions are to be reviewed to determine if they are appropriate, and if they have a significant impact on the results.

Members of the Peer Review Panel

- **Dr. Ivan Catton:** Professor at the University of California, Los Angeles, California. Prof. Catton is an internationally recognized expert in thermal-hydraulics, and has served as a member of the ACRS for the NRC.
- **Dr. David Johnson:** Vice President of ABS Consulting Inc., Irvine, California. Dr. Johnson is an internationally recognized expert in PRA. He is involved in major risk studies and in the use of those studies to support decision-making.
- **Dr. Thomas E. Murley:** The chair of this peer review panel is a former Director of the NRC's Office of Nuclear Reactor Regulation (NRR). Dr. Murley played a key role in regulating the operation of nuclear power plants for many years in comprehensive, high-level, broad-scope management of programs on water-cooled nuclear reactor power plants' safety and risk assessments.
- **Dr. Upendra Rohatgi:** Researcher at the U.S. Department of Energy's Brookhaven National Laboratory, Upton, NY. Dr. Rohatgi has been extensively involved in the development of thermal-hydraulic computer codes development for nuclear power plant applications. In the mid-1980's he reviewed the thermal-hydraulic analyses performed for two of the plants analyzed during the development of the current version of the PTS Rule.
- **Mr. Helmut Schulz:** Head of Department of Structural integrity of Components at GRS (Gesellschaft fuer Anlagen-und Reaktorsicherheit), Cologne, Germany. Mr. Schulz has been involved as a senior manager directing the development of PFM methodologies and managing various international cooperative research projects concerning fracture mechanics under the auspices of the CSNI (Committee on Safety of Nuclear Installations) and the Nuclear Energy Agency (NEA) of Organization for Economic and Development (OECD) in Europe.
- **Dr. Eric vanWalle:** Head of the Reactor Materials Research Department, Belgian Nuclear Research Center (SCK-CEN), Mol, Belgium. Dr. vanWalle is extensively involved in irradiation embrittlement characterization of RPV materials, and in various International Atomic Energy Agency (IAEA) and OECD/NEA cooperative research projects in fracture mechanics related to ensuring the structural integrity of nuclear power plants.

Peer Review Meetings

Three meetings were held with the peer reviewers to provide face-to-face direct interactions with the investigators in the each of the key subject areas. The industry observers were invited to attend these meetings. The first meeting (public) was held during November 17 through 19, 2003, in which the developed methodology was discussed and potentials shortcomings in specific areas were pointed out. The second meeting focused on TH methodology and uncertainty evaluation was held on April 26-27, 2004. The TH methodology review meeting involved the two TH reviewers and the chair of the review panel. In the third review meeting (public), held during May 10-11, 2004, the results obtained using the developed methodologies were discussed, and plans for completion of the remaining analyses were presented. Following the May 2004 peer review meeting, a few additional refinements in the methodology were completed and their effect on the results were assessed.

Process for Obtaining and Addressing Review Comments

Prior to the first peer review meeting detailed information about the developed methodology was provided to the review panel. During December 2003 through February 2004, each of the peer reviewers provided written comments on their subject areas of specialization and also on the overall methodology. These comments and staff response are provided following this page, as comments numbered 1 through 76 (inclusive). These responses were provided to the peer reviewers along with all of the reports detailed in Figure 4.1 that appears in the main body of this report. Following their review of these reports (in general) and the staff's response to their comments (in specific) each of the reviewers provided a letter. These letters appear at the end of this Appendix, along with staff responses (where appropriate).

References

1. Mario V. Bonaca, Chairman, Advisory Committee on Reactor Safeguards (ACRS), U.S. Nuclear Regulatory Commission, "Pressurized Thermal Shock (PTS) Reevaluation Project: Technical Bases for Potential Revision to PTS Screening Criteria," February 21, 2003.
2. William D. Travers, Executive Director for Operations, U.S. Nuclear Regulatory Commission, "Pressurized Thermal Shock (PTS) Reevaluation Project: Technical Bases for Potential Revision to PTS Screening Criteria," March 28, 2003.
3. "ASME Standard RA-S-2002, "Standard for Probabilistic Risk Assessment for Nuclear Power Plant Applications," An American National Standard, The American Society of Mechanical Engineers, April 5, 2002."

Peer Review Comments and Staff Responses

The remainder of this appendix lists each written comment received from the peer reviewers, along with the related staff responses, or provides references to specific reports where the responses can be found.

The following pages provide the staff's responses to comments made by the reviewers following meetings held in December 2003 and February 2004.

Reply to Reviewer Comment #1

Comment made by: Murley

Reply by: MEB

Comment: It appears that the NRC staff is intending to keep the current form of 10 CFR 50.61, which sets minimum fracture toughness requirements on PWR pressure vessels by means of screening limits on the surrogate parameter RT_{NDT}. Any difficulties in implementing the rule would only arise if the screening criteria were approached and the plant's licensee was not able to demonstrate that practicable flux reduction programs would prevent RT_{PTS} from exceeding the screening criteria at the end of life of the reactor. In such an event the licensee would be required to carry out a detailed PTS risk analysis. Alternatively, the licensee could choose to thermally anneal the vessel under the requirements of 10 CFR 50.61. If this form of the PTS Rule is maintained, clearly the NRC must issue revised detailed guidance on how a licensee's PTS risk analysis is to be carried out.

Staff Response: These issues will be addressed in rulemaking. Answers to this question cannot be provided until the actual rule is structured by NRR. Any response prior to that time would be premature.

Comment made by: Murley

Reply by: PRAB

Comment: There is no discussion of events like the 1978 Rancho Seco overcooling event, where the vessel was cooled from 582°F to 285°F in slightly over 1 hour, while reactor pressure was about 2000 psi. A control system error reduced main feedwater flow, causing the reactor to trip on high pressure. The auxiliary feedwater started and the resultant primary system cooldown and pressure drop actuated the high-pressure injection pumps and all auxiliary feedwater pumps. Because their instruments had failed, the operators maintained HPI and aux feed for one hour. While the proximate cause of this event has been corrected, there may well be similar events that should be considered.

Given that the event did actually happen, and that analyses of that era indicated it was a major safety concern, what's changed so radically since then that we now think such events are not a safety concern, and that mainly primary system breaks cause the large majority of PTS risk?

Staff Response: **Short Event Description:** A shorted direct current (DC) power supply caused loss of power to the plant's non-nuclear instrumentation (NNI), which caused the loss of most control room instrumentation and the generation of erroneous signals to the plant's integrated control system (ICS). The ICS reduced main feedwater (MFW), causing the reactor to trip on high pressure. The cooldown was initiated when feedwater was readmitted to one steam generator (SG) by the ICS; auxiliary feedwater (AFW) was restored. The cooldown caused system pressure to drop to the setpoint (1600 psig) for safety features actuation, which started the high-pressure injection (HPI) pumps and AFW to both SGs. HPI flow restored pressure to 2000 psig. With control room instrumentation either unavailable or suspect for 1 hour and 10 minutes (until NNI power was restored), operators continued AFW and MFW to the SGs, while maintaining reactor coolant system (RCS) pressure with the HPI pumps. Analyses of the event (by NRC/RES) indicated that, had the event happened later in the plant's life, and if a 1-in. flaw had existed in the vessel, the vessel would have failed.

Since that event (and its analysis), several things have changed that tend to reduce the perceived importance of the event:

- modifications to the ICS (allowing it to more effectively deal with similar occurrences)
- redesign of the control room instrumentation to provide operators with more reliable level indication (allowing them to more appropriately respond to the event)
- improved operator training and procedures to deal with potential overcooling scenarios (allowing them to better recognize and respond to similar events and thereby affecting human error probability estimates)

- Fracture mechanics calculations can now be based on current knowledge of thermal-hydraulic (TH) conditions, materials composition, flaw density, and flaw propagation. These improvements allow a more realistic estimation of the probability that such an event could result in a through-wall-crack.

Each of the above contributes to the reduction in importance of events similar to the Rancho Seco event. This does not imply that such an event is impossible, just that its perceived importance has been reduced. For example, as part of the human reliability analysis (HRA) performed for Oconee, Beaver Valley, and Palisades, the distributions associated with "operator fails to control feedwater" (MFW or AFW) were derived considering how lack of level indication (or false level indication) might affect the operator's response. For many situations, this enhanced HRA resulted in substantial "credit" for operator response (when all factors were considered), thereby reducing the importance of such events. These lower operator failure probabilities, in conjunction with the other three items, tend to reduce the importance of events involving SG overcooling. In addition, the current integrated probabilistic risk assessment PRA/HRA, TH, and fracture mechanics calculations indicate that events involving primary system breaks are important. From a PRA/HRA perspective, there is very little the operators can do to "minimize" the cooldown associated with primary breaks involving medium and large break loss-of-coolant accidents since primary injection is required to prevent core damage. For those primary breaks involving stuck-open valves that suddenly reclose, typically the operators have very little time to perform actions that will minimize the rapid increase in pressure. This limited time translates to minimum "credit" for controlling pressure; thus, the increased importance of such events. All of these factors in combination tend to reduce the importance of SG overfeeds and increase the importance of primary system breaks.

Although the exact "Rancho Seco" event was not analyzed as part of the present PTS project, similar sequences were analyzed for Oconee (a B&W plant and, thus, the plant most similar to Rancho Seco). The Oconee sequences that were most similar to the Rancho Seco (RC) event involved the following:

(1) a reactor/turbine trip

(2) one or two stuck-open relief valves on one or two steam generators (possibly a little worse than the RC event)

(3) MFW and AFW continuing to provide water to the steam generators

(4) high-pressure injection such that primary pressure reaches the pressurizer safety relief set point (again possibly a little worse than the RC event)

Recent estimates of the conditional probability of failure (CPF) estimated using FAVOR (the probabilistic fracture mechanics code currently being used by the staff and its contractors) indicates that the CPFs for those bins were zero for all of the above events, even for the Oconee RPV artificially assumed to have been embrittled to the equivalent of 1000 EFPY of operation.

Thus, there are the initial reasons given first (above) to argue that the Rancho Seco event would not be likely, given the changes in plant design and operation that have occurred since the RS event occurred. Moreover, even if it an event similar to Rancho Seco were to occur, today's fracture mechanics calculations indicate that it ***would not*** fail the vessel, even for conditions of embrittlement that are not considered likely to occur within 60 years of operation.

Reply to Reviewer Comment #3

Comment made by: Murley

Reply by: PRAB

Comment: I believe the dominant contributor to TWCF for external events, a small LOCA caused by a seismic event, warrants a more realistic analysis to judge the conservatism, if any, in the presumed bounding TWCF estimate of 3E-8 per year.

Staff Response: Section 9.4 of this document provides more backup and clearer tables on the small LOCA analysis, including consideration of a seismically induced LOCA. The main fact is that the external event analyses are done conservatively, for the reasons noted in Chapter 9 of NUREG-1806. Given that those results are conservative, the conclusion is that the total PTS TWCF can be approximated using only the internal event results. The amount of conservatism introduced into the overall process by this analysis is acceptable (i.e., it doesn't change the final result to such an extent that relaxation of the PTS Rule cannot be considered). For that reason, and because further external events analyses would have to be detailed and plant-specific (i.e., time consuming and expensive), the staff has not made such analyses.

Reply to Reviewer Comment #4

Comment made by:	Murley
Reply by:	PRAB
Comment:	This PTS reevaluation did not consider event sequences involving external flooding of the reactor pressure vessel cavity, which would overcool the outside vessel wall and which occurred at Indian Point several years ago.

Staff Response:

A review of the Indian Point 2 evaluation of the reactor cavity flooding event identifies that there is a much lower risk associated with the external cooling of the reactor vessel versus the events that have been evaluated as part of this project. For example, as water moved up toward the Indian Point RV, it would contact the insulation first. As the water contacts the area between the vessel and insulation, the hot air would flash the water to steam. The steam would be at 212°F. At equilibrium, an estimated temperature drop of 50°F between the steam blanket and the vessel surface would exist, which would leave the vessel outside surface temperature at an estimated 262°F. This warm temperature along with the fact that the material on the outside of the vessel wall is much less embrittled considerably reduces the risk of a transient producing a through-wall crack in the vessel. Additionally, this information indicates that external cooling produces a transient that, at worst, is only as severe as a main steam line break. The results presented in Chapter 8 of this report show MSLB transients to be much less severe than any primary side transient.

Considering all of these factors the risk of through-wall cracking initiated by external cooling of the vessel is believed to be sufficiently small that it can be appropriately ignored in this study.

Comment made by: Murley

Reply by: SMSAB

Comment: It seems to me that the state of the art of thermal-hydraulics (TH) represented in this PTS analysis has not advanced much since the early 1980s.

Staff Response: The capability to analyze PTS scenarios has undergone a revolutionary change since the first PTS study. The 1980s study was greatly limited by the ability to analyze different scenarios. Enormous advances in analysis tools (automated processes and plotting and data extraction routines) also have occurred. These tools lead to more comprehensive analyses, extensive use of sensitivity studies, better communication and sharing of data, and more effective reporting of results. Computing efficiency has increased by orders of magnitude due to increased speed and reduced cost.

The RELAP5 code has been improved as well. The first PTS study was performed during the early 1980's. In this study, RELAP5 thermal-hydraulic calculations were performed for the Oconee Unit 1 plant and for the H. B. Robinson Unit 2 plant. The Oconee calculations were performed with RELAP5/MOD1.5 (circa 1982) and the H. B. Robinson calculations were performed with RELAP5/MOD1.6 (circa 1984). The results of these calculations were documented in a series of NUREG/CR reports, including NUREG/CR-3761 and NUREG/CR-3977 for the Oconee plant and NUREG/CR-3935 for the H. B. Robinson plant.

The RELAP5 calculations performed for the PTS Reevaluation Project are being performed using RELAP5/MOD3.2.2Gamma, which was released in 1999. The changes in the RELAP5 code over the intervening 20-year period between the PTS studies have been extensive. These changes include a revised treatment of non-equilibrium behavior models, including wall heat transfer models and also coupling of the wall heat transfer and vapor generation models. Interphase frictions models were revised, including incorporation of a new interphase drag model for the vertical bubbly and slug flow regimes. A general cross-flow modeling capability was installed, allowing cross-flow connections to be made between most types of components and among the cell faces on those components.

Other changes were implemented as a result of the code assessments related to the RELAP5 analysis for AP600. The Henry-Fauske critical flow model was added to the code, providing a standard-reference critical flow model upon which code calculations are based. Changes were made in code numerics that greatly reduced recirculation flows within model regions nodalized with a multidimensional approach. A mechanistic interphase heat transfer model was implemented that includes the effects of noncondensible gases; this change greatly improved the simulation of condensation, preventing erratic behavior and code execution failures. This change is particularly important for situations where the plant accumulators empty and nitrogen is discharged into the reactor coolant system (a situation that typically led to code execution failure at the time of the first PTS study).

For PTS Reevaluation Project analysis, no major changes were made from the RELAP5 plant input modeling approach used in the prior PTS study. With only a few exceptions, the plant input models use the same nodalization schemes as before. Those nodalization schemes reflect plant modeling recommendations and guidance for the general modeling of plant transients, which evolved over years of RELAP4 and RELAP5 experimental assessments and plant applications preceding the first PTS study. However, capabilities in RELAP5/MOD3.2.2Gamma are utilized including renodalization of the reactor vessel downcomer (using the general cross-flow modeling capability), conversion of the vessel/hot and cold leg connections and the hot leg-to-pressurizer surge line connection to the cross-flow format, and addition of junction hydraulic diameter input data as required by the conversion of the code to junction-based interphase drag.

Current computer calculation speeds and data storage capabilities are at levels unimagined at the time of the prior PTS study, allowing the number of transients that can be reasonably evaluated directly using RELAP5 to be expanded by more than an order of magnitude. In the prior PTS study, budget and schedule considerations limited the number of transients evaluated per plant to about 10 to 15. The number of transients used to characterize the risk of vessel failure in the current PTS Reevaluation Project is over 500.

<u>Comment</u>: The PTS PIRT (phenomena identification and ranking table) considers mostly system parameters, like flows and pressures, but little analysis of conditions in the downcomer

<u>Staff Response</u>: Several items are included that focus on the downcomer. First, the PIRT includes jet behavior, flow distribution and mixing in cold leg. This includes several related phenomena under the heading of fluid-fluid thermal mixing. It includes the mixing in the ECC injection line before the flow reaches the cold leg, mixing of the ECC jet where it enters the cold leg, stratification in the cold leg, mixing of the stratified flow within the cold leg as it moves towards the vessel, and backflow of ECC liquid from the upper downcomer towards the RCP and loop seal.

Second, the PIRT also includes jet behavior, flow distribution and mixing in the downcomer. This comprises several closely related processes under the heading of fluid-fluid thermal mixing. It includes a number of flow and mixing phenomena such as whether the fluid stream tends to hug the core barrel or vessel wall, mixing as the flow enters the downcomer from the cold leg and turns from horizontal to vertical, and plume decay.

Third, the PIRT considers convective heat transfer in the downcomer. The experimental data base was reviewed with data identified from UPTF, APEX, and Creare with which the modeling of downcomer flows and heat transfer were assessed in RELAP.

Fourth, it included in-vessel buoyancy driven natural circulation flows through the upper plenum-upper downcomer bypass, and B&W vent valves, level formation in the downcomer, and condensation in the cold leg during ECC injection. Other aspects of downcomer conditions are mainly determined by system parameters and how they determine the RCS pressure and energy distribution within the RCS.

Comment: I did not see any comparisons of calculations with measured vessel wall temperatures.

Staff Response: As a result of peer review comments, these comparisons were carried out more recently. Data were identified from UPTF, APEX, and Creare with which RELAP5 was compared (NUREG-1809). Integrated assessment was performed comparing RELAP5 predictions of flows and wall-to-fluid heat transfer. The assessment showed that RELAP5 was realistic or conservative with respect to the experimental data. No nonconservatisms were identified.

Comment made by: Murley

Reply by: SMSAB

Comment: The report states that "downcomer heat transfer coefficient variation has little contribution to PTS risk uncertainty". This may be because the uncertainty analysis only considered variations of ±30% from nominal values. What is the basis for that limitation? What is the effect on TWCF of larger uncertainties in heat transfer coefficients?

Staff Response: Since the April 2004 draft of this report on which Dr. Murley commented, additional work was performed addressing convective heat transfer. This work is reported in NUREG-1809 (see Chapter 5 and Appendices E through H or this document. This work shows that the effect of uncertainty in $h_{dc}(t)$ is similar to the uncertainty in temperature because these two parameters are part of one and the same question: that being the impact of the uncertainty on q", which is defined as follows:

$$q''_{dc}(t) = h_{dc}\,(T_w - T_f)$$

Comparisons of RELAP5 with integral experimental data from UPTF and APEX-CE under conditions of loop flow stagnation show that RELAP's prediction of $h_{dc}(t)$ is realistic or conservative; no nonconservatisms were identified. Here, the word "realistic" is used to mean that the value $h_{dc}(t)$ are within ≈20% of measured values. On this basis variations of $h_{dc}(t)$ above ±30% cannot be viewed as credible, so there is no relevance in assessing the effect of such variations on TWCF

Comment made by: Murley

Reply by: SMSAB

Comment: The RELAP 5 calculated flow in the 2 X 4 plants is not predicted realistically. The flow is sometimes negative in one cold leg, while positive in the other cold leg. Similarly, in the Palisades simulation recirculation flow was seen in the axial direction in parallel downcomer flow channels. Are these calculational anomalies understood? The use of artificial check valves in the calculational models is troubling to me.

Staff Response: Under conditions of loop flow stagnation, the pressure driving forces and buoyancy driving forces for flow are small. Such conditions do not exist universally for all calculations, but rather, appear only in certain circumstances. Numerical solutions to the momentum equation can be unstable. The frictional and form loss resistances to flow are also small. In a systems code, turbulent viscous dissipation is not, and cannot be, represented. This was a point of emphasis at the start of the AP600 design certification review in 1990.

The cause of the numerical flows is basically understood. The numerical initiator indicated that flow first begins due to round-off errors (in the last digitally-stored significant digits) in the pressure solutions at the ends of the identical pipes. *The numerical initiator is therefore judged to be unavoidable when using a digital thermal-hydraulic systems code.* In a 2 x 4 arrangement, two cold legs connect to the same volume at either end, and are therefore identical. A high resistance in the negative flow direction (vessel towards steam generator) was used in the RCP to provide damping to the solution to avoid numerical flows. This avoids mixing that would otherwise occur, and causes downcomer temperatures to be colder. Therefore, this numerical effect imposes a conservative bias in the results. More information on RELAP5 numerical issues, and on the appropriateness of our approach, can be found in Section 6.3.2 of this report and in Appendix C to NUREG-1809.

Comment made by: Murley

Reply by: MEB

Comment: In something as complex as a PFM calculation it is necessary to compare calculations with as broad a range of experimental data as practical. For this reason, it is important to complete the Validation and Verification report on the FAVOR code well before the peer review panel's final meeting in April. Similarly, the report on PFM sensitivity studies scheduled for March 2004 will be important for panel review.

Staff Response: FAVOR verification and validation (V&V) involves assuring that the software meets the requirements stated in the FAVOR theory manual. A report detailing FAVOR code V&V is available [***Malik***]. However, it should be noted that this report does not concern comparison with experimental data. Experimental data is addressed by [***EricksonKirk-PFM***]. Additionally, in Appendix.A to [***EricksonKirk-PFM***], we predict the outcome of scaled PTS experiments conducted on scaled pressure vessels.

Reply to Reviewer Comment #9

Comment made by: Murley

Reply by: PRAB

Comment: Much less study of the consequences of RPV failure accidents has been done than is the case for core damage accidents resulting from undercooling or ATWS events. As a result the question arises whether vessel failure accidents could lead to especially large early release scenarios. In particular the ACRS has raised the issue of potential Large Early Release Frequency (LERF) source terms from air oxidation of fuel in some of the most severe (and unlikely) RPV failure scenarios. I do not think it would be a wise use of resources to mount a substantial research effort to try to answer all the questions surrounding air oxidation source terms. Perhaps a modest expert elicitation task might produce a consensus on bounding consequences of such scenarios. In any event this PTS project is not the place to revise the Commission's policy on LERF guideline.

Staff Response: This was discussed at the PTS peer review group meeting on May 10–12, 2004. The NRC staff agrees with Dr. Murley that the PTS project is not the best place to establish the Commission's policy regarding LERF.

Reply to Reviewer Comment #10

Comment made by: Murley

Reply by: PRAB

Comment: The staff makes a reasonable case that the conditional probability of a large early release of radioactivity, given a PTS-induced RPV failure, is small (less that 0.1) to extremely small (much less than 0.01). Based on their largely qualitative analyses, the staff suggests an acceptance criterion of TWCF = 10^{-6}/ry or less. I expect that the NRC staff will address this issue in the planned Engineering Summary Report and Executive Summary Report scheduled for April. I plan to comment further on this issue in my final report after further review.

Staff Response: The staff discusses this subject in Chapter 10 of this report. The TWCF of 10^{-6}/ry was developed using current NRC guidance for LERF in RG 1.174.

Reply to Reviewer Comment #11

Comment made by: Murley

Reply by: MEB & NRR

Comment: What are the regulatory requirements for a plant that has suffered a severe overcooling event where the vessel did not have a thru-wall crack and no outward sign of damage but may have suffered a crack initiation that subsequently arrested in the vessel wall?

Staff Response: As this question addresses current regulatory requirements members of NRR staff prepared a response that can be found in the NRC's Agencywide Documents Access and Management System (ADAMS), under Accession #ML41700384. The staff's response included the following information.

An overcooling event would violate the facility pressure-temperature limits by amounts dependent on the specific event. The facility's technical specifications will identify the specific actions the licensee is required to take in the event of a violation of these limits. From the standard technical specifications (and we expect that all plants would have similar provisions in their technical specifications), licensees are required to restore their pressure and temperature to within established limits within 30 minutes, and determine if the reactor coolant system is acceptable for continued operation within 72 hours. In addition, such a violation would invoke the reporting requirements given in 10 CFR 50.72 and/or 10 CFR 50.73, which would ensure that the NRC would be notified of the event. While obtaining information to respond to your question, the staff has discovered that the technical specifications for one of the plants that shows high PTS sensitivity do not have the latter provision. The Office of Nuclear Reactor Regulation (NRR) will discuss with this licensee the inconsistency between their custom technical specifications and the standard technical specifications and seek to resolve the inconsistency.

Given the occurrence of any overcooling event that violates facility operating limits, an evaluation for continued operation must verify that the reactor coolant pressure boundary integrity remains acceptable and must be completed if continued operation is desired. Several methods may be used, including comparison with pre-analyzed transients in the stress analyses, new analyses, or inspection of the components. The American Society of Mechanical Engineers (ASME) Code Section XI, Appendix E, "Evaluation of Unanticipated Operating Events," may be used to support the evaluation. If the acceptance criteria given in Appendix E are satisfied, the staff would conclude that the facility is unlikely to have suffered a crack initiation-arrest event during the observed severe overcooling transient (i.e., the pre-existing flaw population would be unaffected by the event) and that continued operation of the facility is acceptable. If the analysis specified in this appendix does not justify continued operation, then the appendix states that additional analyses or other actions shall be taken to assure that acceptable margins of safety will be maintained. It is likely that the other actions would involve an inspection of relevant portions of the RPV.

Based on past experience with a licensee who experienced a modest overcooling event that exceeded the facility's pressure-temperature limit curve, it is likely that a licensee would evaluate the structural integrity of any identified flaws in accordance with ASME Section XI, Appendix A, "Analysis of Flaws," to determine whether the flaw(s) could have grown during the overcooling transient. Although such an analysis would not be required by NRC regulations, the NRC would review this information in the context of the licensee's determination that the RPV was acceptable to return to service.

Severe overcooling events are, based on facility operating experience, infrequent events. Given that a severe overcooling event is the result of an unanticipated plant condition, 10 CFR Part 50, Appendix B, would require that the root cause of the event be evaluated and action be taken to mitigate the potential for a second severe overcooling event due to the same root cause. Once this known susceptibility is addressed, the affected facility is as unlikely as any other similarly designed facility to suffer another severe overcooling event. If the root cause evaluation points to a generic condition which could make other facilities subject to similar events, then NRC staff would evaluate the need for a generic communication to the industry on the topic, or other regulatory actions.

Reply to Reviewer Comment #12

Comment made by: Murley

Reply by: MEB & NRR

Comment: How would the PTS risks change if such a cracked vessel went back into service?

Staff Response: Again, the staff's response can be found at ADAMS Accession #ML41700384. We believe that the evaluations in accordance with ASME Section XI Appendices A and E (or similar evaluations) and the inspections (if deemed necessary) described above will preclude continued operation with unanticipated flaws as a result of a severe overcooling transient. Again, if a licensee were to fail to pass the screening analysis in ASME Section XI, Appendix E, we would expect that inspections capable of finding any flaw (with a probability of detection near unity), which might have resulted from a crack initiation-arrest event would be conducted before returning the vessel to service. Assuming that an arrested flaw has not been detected, risk may increase depending upon the type of transient(s) to which the arrested flaw is later subjected. The specific thermal-hydraulic characteristics of any transient which would occur after the crack initiation-arrest event would be critical in determining the quantitative risk increase, if any, associated with the arrested flaw. We believe that the analysis and inspection activities described above provide reasonable assurance that such an undetected flaw would be very unlikely to occur. Hence, based on the combination of events that would have to occur in order to lead to potential vessel failure, we qualitatively believe that the impact on overall plant risk of the scenario that you proposed is very low.

Comment made by: Murley

Reply by: MEB & NRR

Comment: What criteria would NRC use to judge whether a complete inspection of the vessel was needed after a severe overcooling event? What inspection techniques would be required and what would be the scope of such inspections?

Staff Response: As this question addresses current regulatory requirements members of NRR staff prepared a response that can be found at ADAMS Accession #ML41700384. The staff's response included the following information.

The "other actions" specified in ASME Section XI, Appendix E (if the analysis does not justify subsequent operation) would likely be an inspection of the RPV. The determination of the need for inspection, as well as the inspection technique and inspection scope, would be based on (1) the known embrittlement level of the vessel in question, (2) the severity of the overcooling event (i.e., the cool down rate of the transient and the duration time of the transient), (3) the results of prior licensee in-service inspections of the vessel, and (4) the results of the licensee's vessel structural integrity evaluation described above.

Reply to Reviewer Comment #14

Comment made by: Murley

Reply by: MEB

Comment: the explanation for how the original screening criteria were selected is not only confusing (especially Figure 1.1) but it is incorrect as well. .... In SECY 82-465 the staff stated quite clearly that "the large uncertainties in probabilistic PTS evaluations at the present time (1982) have led the staff to use them to estimate the level of safety rather than attempt to derive licensing requirements directly from the probabilistic results....the NRC staff recommends that the PTS criteria-screening or otherwise-should not be determined by where these curves cross some acceptable value of risk". It is important to correct these types of errors, not because they affect the substance of the technical analyses but because they can undermine the credibility of the entire effort. This particular error, for example, would give a reader the impression that the current PTS reevaluation is using the same regulatory rationale as in 1982 with better data and calculations, whereas the truth is that the proposed approach to the PTS Rule change represents a fundamental change in the amount and use of probabilistic information in the formulation of new screening criteria.

Staff Response: See corrected explanation in Section 2.3 of this report.

Comment made by: Schulz

Reply by: PRAB

Comment: Main focus of the event sequences is a range of power operation from zero power hot stand-by up to 100% power. Sequences which may result out of malfunctions or wrong operator actions during start-up up to zero power hot stand-by cool down from zero power to residual heat removal and test conditions are not included in this study. The reviewer feels that additional justification is needed in this respect. For example, in some Russian units we have seen sequences happened that during a pressure test with the core loaded the primary safety valve opened inadvertently followed by a full ECCS injection at a vessel temperature below operating temperature.

Staff Response: This comment involves both low-temperature over-pressure (LTOP) situations, and situations where the event starts with nominal temperature and pressure conditions but at "zero power" (i.e., at "hot, zero power" (HZP), or nearly so, conditions). Regarding LTOP situations, LTOP involves cold conditions in a primary system that's closed (i.e., that could be accidentally pressurized), the operation of shutdown cooling systems, and the operators' use of procedures appropriate for those conditions (as opposed to normal operating procedures), that is, it involves system and operator operations just before or after shutdowns such as refueling. Analyses for those conditions are quite different from the PTS analyses we have performed for this study, and thus are outside the scope of the PTS analyses. Separate programs have been conducted to deal with such LTOP conditions. Regarding HZP (or nearly so) conditions, our assumption that about 2% of the time (per year) is spent at hot zero power should cover those situations that are "nearly HZP" as well, since our plant analyses suggest that plants are at HZP more in the range of 1% to 1½% of the year, so by rounding up to 2%, we believe we have bounded any "near-HZP" transition states as well. Both of these issues were discussed and (the staff believes) satisfactorily resolved at the peer review meeting on May 10–12, 2004.

Reply to Reviewer Comment #16

Comment made by: Schulz

Reply by: PRAB

Comment: In principle, only cool down of the vessel from the inside has been investigated. It may be useful to document that severe cool down scenarios from the outside of the vessel as we have seen as a precursor for Indian Point some decades ago can be ruled out. The reviewer is not so familiar with PWR system designs for US units to judge if this is a relevant question at all.

Staff Response: Please refer to the response to Comment #4.

Reply to Reviewer Comment #17

Comment made by: Schulz

Reply by: SMSAB

Comment: Looking to the different system functions all possible combinations are exhaustedly studied for the operational modes being investigated. It is the understanding of the reviewer that the thermo-hydraulic analysis assume in principle that the component internals function as designed. We have seen already in the past experience that degradation of internals of the reactor pressure vessel and steam generator take place. Some of the transients being investigated would impose severe loads on internals. Therefore it may be worthwhile to investigate the likelihood, if consecutive failures at the internals could produce aggravated loading conditions for the reactor pressure vessel.

Staff Response: Vessel internals are designed for blowdown loads from a large-break LOCA, however, the comment suggests that materials problems may degrade the integrity of vessel internals over time. The question, then, is whether internals may experience failures as a direct consequence of a PTS event that would lead to more adverse downcomer conditions. The only way this could happen would be if such a failure led to lower temperatures in the downcomer, since pressure gradients are not a factor. We are unable to identify any postulated failure of internals that could lower the downcomer temperature. The type of failure that would be required would be if all the vent valves in a B&W design failed shut. Such a failure of multiple parallel check valves is implausible. Any failure that increased bypass flow between the downcomer and upper plenum would tend to increase downcomer temperature.

Comment made by: Schulz

Reply by: SMSAB

Comment: As discussed already by other colleagues, the thermo-hydraulic calculation using RELAP-5 produce more or less mean temperature values in the downcomer at each time step. From the analysis results I have seen in the course of several safety assessments I conclude that nonuniform temperature distribution in the downcomer produce non symmetric loading conditions which have at least an impact on crack initiation of surface breaking flaws. The time of crack initiation and the orientation of flaws which would initiate would be different from analysis results using purely symmetric cool down. It is difficult to judge for the reviewer if significant differences would result between nonuniform and uniform loading conditions for embedded flaws and cracks being extended to a considerable fraction of the vessel wall thickness. As it has been seen in the UPTF test the nonuniform condition caused by local mixing are not stable in space so model assumptions using the most pronounced nonuniform temperatures may be overly conservative for flaw locations beneath the surface. I assume that this aspect will be discussed in more detail by other colleagues.

Staff Response: The first point to consider is whether nonuniform temperature exists to begin with. Review of integral system experimental data from LOFT, ROSA, and APEX-CE do not show the existence of thermal plumes. Full-scale data separate effect from UPTF show limited temperature nonuniformity (~20°C at most, and often less). A more complete discussion is found in [***Bessette***].

Reply to Reviewer Comment #19

Comment made by: Schulz

Reply by: MEB

Comment: Considerable effort has gone to the development of a more realistic flaw model by enlarging the experimental data sources. With the material available it is difficult to judge to what extent the sample material is representative for the whole set of vessels where the revised PTS Rule would be applicable. The reviewer is not familiar enough with the fabrication practice in the 1960s and 70s as well as the differences in practice between the different manufacturers. To my knowledge the ultrasonic inspections during manufacturing in the 1960s and 70s were largely voluntarily and not required by the code at that time. The in-service inspections following ASME XI are basically addressing welds. Concurrent with previous discussions (SECY/82/465) the reviewer would assume that a revised PTS Rule would also address the requirements on ISI and NDE qualification. It is the view of the reviewer that a flaw model as outlined should only be used under conditions such as:

1. Applicability check of the flaw density distribution for the pressure vessel under consideration including similarity check of fabrication practice.
2. Applicability check of the flaw density distribution supported by non-destructive testing results for the near core region for weld and base material either using existing inspecting records or establishing a new finger print. In case only embedded flaws are used in the fracture mechanics analysis the necessary reliability of NDE to rule out surface breaking flaws may not be achievable by applying only ultrasonic methods, e. g. looking to one of the most recent exercises (NESC-1).
3. It is the understanding of the reviewer that the flaw model is basically addressing remaining manufacturing defects. Although the operating experience with PWR vessels is judged to be favorable by the technical community it has to be remembered that the inspection of the cladding is not required by the ASME XI and being applied only in a few countries. Therefore the present view of the technical community may not be adequately based on inspection records.

Staff Response: The applicability of the flaw distribution adopted in FAVOR to PWRs in general is addressed in both Appendix C of this document as well as in [***Simonen***]. It is the view of the staff that the flaw distribution adopted in FAVOR is applicable to PWRs in general, in part because of the known conservatisms that are part of the flaw model.

Details of how/if ISI/NDE requirements are incorporated into any future revision of the PTS Rule will be addressed as part of rulemaking

Reply to Reviewer Comment #20

Comment made by: Schulz

Reply by: MEB

Comment: It is the understanding of the reviewer that the flaw density distribution and material property distributions are used as independent variables. To my knowledge this is common practice but may not reflect the real situation for all kind of defects. From the experience of the past we have seen that crack like defects are governed to some extent by unfavorable material properties at certain locations. The reviewer admits there is no reliable data base to establish a correlation factor but still the sensitivity may be addressed in a parametric study assuming certain correlation factors.

Staff Response: We agree with the Dr. Schultz's assessment that *"there is no reliable data base to establish a correlation factor."* Absent such information, there is no credible basis for the sensitivity study suggested.

Reply to Reviewer Comment #21

Comment made by: Schulz

Reply by: MEB

Comment: The use of correlations between types of material tests (Charpy, K_{IC}, T_o) which characterize the toughness behavior was and will be a topic of discussion within the technical community. The basic principle of the presented methodology to provide means of assessing PTS risk without requiring licensees to make more measurements on the vessel material seems from my view a difficult regulatory position. The proof that the range of uncertainties is small enough for an individual vessel should remain with a licensee. Complementary irradiation programs which would produce additional fracture toughness data are not judged to be an unnecessary burden for a considerable extension of the life time.

Staff Response: Specific regulatory requirements regarding the data that must be supplied by the licensee for vessel specific materials will be established as part of rulemaking, if rulemaking is undertaken.

It should be noted that the uncertainty in material data (Cu, Ni, P, $RT_{NDT(u)}$, yield strength, upper shelf energy, and so on) assumed by FAVOR is based on generic information and, therefore, is larger than the uncertainty associated with any plant-specific materials. As such, the treatment of material uncertainty that underlies the PTS screening criteria recommended in Chapter 11 of this document is conservative relative to the uncertainty of plant-specific materials. See Appendix D of [***EricksonKirk-PFM***] and Section 4.2.2.2 of [***EricksonKirk-SS***] for a detailed discussion.

Comment made by: Schulz

Reply by: MEB

Comment: For the fracture mechanics approach being used the status of validation/verification is well demonstrated for crack initiation and limited stable or unstable crack extension. Although present safety standards or codes do allow the application of crack arrest for multiple initiating events in principle, the supporting experiments are very limited. Furthermore, to the knowledge of the reviewer analysis being performed on such tests (for example NKS test at the MPA Stuttgart) were not able to predict consistently re-initiation and multiple arrest conditions. To evaluate the need to address this issue more deeply, it would be helpful to know if multiple initiation and arrest conditions are really connected to the scenarios being investigated or are only treated as theoretical possibility.

Staff Response: Given that a flaw may exist within the wall of nuclear reactor pressure vessel (RPV), it is consistent with U.S. experimental evidence and analytical fracture-mechanics predictions [Cheverton 85a, Cheverton 85b] that the flaw can propagate into the RPV wall by multiple cleavage run-arrest events when the RPV is exposed to hypothetical pressurized thermal-shock (PTS) loads. To address Dr. Schultz's comment, researchers at ORNL composed a detailed response summarizing this U.S. experience. This response appears in Appendix E, which demonstrates that multiple flaw initiation-arrest events are credible for thick-wall cylinders exposed to thermal-shock transients. Additionally, the nature and extent of such fracture behavior can be adequately predicted by careful application of linear elastic fracture mechanics analyses. The information in Appendix E centers on the thermal-shock experiments (TSEs) that were conducted at ORNL in the late 1970s and early 1980s. Because the basic factors driving the fracture behavior in these TSEs are so similar to those for PTS scenarios, multiple fracture run-arrest events are deemed credible for an RPV exposed to PTS transient loads.

Reply to Reviewer Comment #23

Comment made by: Schulz

Reply by: MEB

Comment: It is the understanding of the reviewer that the FAVOR code cannot handle the effect of pressure acting on the crack surfaces. In case of considerable crack extension of surface breaking flaws - either from the beginning or within the course of the crack extension - the stress intensity at the crack tip may be increased by considering the additional load acting on the surfaces on extended cracks. This may happen in depressurisation scenarios. Although this systematic effect is estimated to be not very big (10-15% increase of K_I) it may well have a significant contribution to the ratio of non-vessel failure versus vessel failure.

Staff Response: FAVOR Version 04.1 was modified to include the effects of crack face pressure in response to Dr. Schultz's comment. See the FAVOR 04.1 theory manual, [***Williams***], and Section 9.2.1.2.6 of this report for details.

Reply to Reviewer Comment #24

Comment made by: Johnson

Reply by: PRAB

Comment: Since the analyses will ultimately support a recommendation from RES to NRR, some consideration needs to be made as to what regulatory guidelines or other standards, if any, are to be followed. I acknowledge that regulatory guidelines and industry standards have evolved significantly since the analyses under review were initiated. In fact, these guidelines and standards are continuing to evolve. Nonetheless, as evidenced by the September 2003 White Paper from Chairman Diaz (COMNJD-03-0002) and the subsequent Commission approval, content and scope of PRA submittals are a key part of achieving "quality" in PRA.

Staff Response: This comment involves both the requirements that future PRAs performed by licensees need to meet, and what requirements and standards the staff and its contractors met in their own PTS work. Regarding requirements for PRAs that licensees may perform in the future, and in particular the extent to which they will need to add PTS sequences to their PRAs, those are policy issues that NRR will need to take into account during the rulemaking action they may undertake, using this RES work as part of its basis (the RES role is to provide the risk-related basis for such a possible action, not to conduct the rulemaking action itself). Regarding requirements and standards the staff and its contractors met, the following observations are provided: (1) this project started in 1999, before the issuance of the full power PRA Standard in 2002, (2) the PRA/HRA project members were aware of the ongoing development of the standard and other documents dealing with PRA quality (e.g., Reg. Guide 1.174) and are familiar with the current Standard, and (3) while no specific review of the analyses have been conducted against the PRA Standard, we believe that, in general, the intent of the Standard has been met.

Reply to Reviewer Comment #25

Comment made by: Johnson

Reply by: PRAB

Comment: Regulatory Guide 1.174 outlines a framework for licensees to follow in formulating risk-informed requests. The purpose of RG 1.174, I believe, is to provide a consistent framework for considering potential plant or procedural changes that could impact risk. The PRA work under review, in contrast, considers a class of scenarios that may or may not be included in the base PRAs. In any event, RG 1.174 provides a framework to consider changes in risk and can be used as a guide, at least for scope and content.

Staff Response: Please refer to the response to Comment #24.

Comment made by: Johnson

Reply by: PRAB

Comment: The PRA analyses estimate or bound the through-wall crack frequency (TWCF) due to thermal shock. RG 1.174, on the other hand, use changes in the core damage frequency (CDF) and large early release frequency (LERF) as surrogates to estimate the impact on public health risk. I believe a discussion of the relation between TWCF and CDF and LERF is warranted. Small increases in LERF will be viewed differently than the same numerical changes in CDF. Does a through-wall crack result in core damage in all cases? One could envision a relative small leak rate from a crack, or a failure that can be mitigated by plant systems. On the other hand, does such a crack result in an "excessive LOCA," or what WASH 1400 called a vessel rupture? Such an event might map directly as a contributor to LERF. Granted these are questions whose answers are unknown, but the analysts need to include a discussion regarding their state of knowledge.

Staff Response: We have assumed that TWCF = CDF. An accident progression event tree (APET) was developed and used to determine the likelihood of events that may lead to LERF. The relationship of CDF to LERF is the subject of Chapter 10, and it was also discussed at the peer review meeting held on May 10–12, 2004 in Rockville Maryland.

Reply to Reviewer Comment #27

Comment made by: Johnson

Reply by: PRAB

Comment: It seems clear that near term PRA submittals will need to meet or discuss the requirements of the ASME Standard (as well as Regulatory Guide 1.200). I strongly suspect that the underlying utility PRAs do not fully meet the Standard. This is probably not a significant point with respect to their technical quality. However, the status of the underlying utility PRAs as well as the RES supported PRA work with respect to the requirements outlined in the Standard and RG 1.200 should be made clear in the submittal to NRR.

Staff Response: Please refer to response to Comment #24.

Reply to Reviewer Comment #28

Comment made by: Johnson

Reply by: PRAB

Comment: Likewise, the ANS Standard governing the conduct of external events has only recently been released and is under review by NRC. How the bounding external events analyses compare to the draft standard should be discussed.

Staff Response: Please refer to response to Comment #24. Additionally, it should be noted that the external event analyses are purposely conservative for the reasons noted in the Section 9.4 of this document. Thus, a direct comparison to any standard is not appropriate, since the analyses are purposely conservative.

Comment made by: Johnson

Reply by: PRAB & MEB

Comment: One of the most interesting results of the current analyses is the recognition that high pressure is not required for damage. In fact, for the PWRs analyzed, large LOCAs represent a key class of initiators that have the potential to result in a scenario involving vessel failure given a sufficient thermal shock. Not being a thermal-hydraulic or fracture mechanic expert, I am tempted to ask, "What about thermal shock scenarios in BWRs?" There may be strong thermal-hydraulic arguments relating to limited cooling rates or strong fracture mechanics arguments relating to the smaller fluence experienced in BWRs that make BWR thermal shock scenarios impossible or of extremely low frequency. If so, the analysis should summarize such arguments as to why BWRs, with thinner vessel walls, are do not have a potential thermal shock concern.

From a systems point of view, I note that for many BWRs, following a rapid depressurization, injection to the vessel is likely to come from the hotwell if the condensate system is available. (In other words, condensate may have a higher discharge pressure than the low pressure coolant injection that draws from the relatively cold condensate storage tank.) Nonetheless, one could postulate a scenario involving loss of feedwater and a stuck-open SRV with HPCI and RCIC injecting water from the condensate storage tank. The point is that if a succinct rationale for excluding BWR scenarios can be given in the analysis, then such an argument should be added to the report. If such scenarios are potentially of interest, then they should be added to the analysis.

Staff Response: SECY-82-465, "Pressurized Thermal Shock (PTS)," November 22, 1982, states on the second page (***emphasis added***), "The PTS issue is a concern only for operating PWRs. Boiling-water reactors (BWRs) are not a significant PTS concern. BWRs operate with a large portion of water inventory inside the pressure vessel at saturated conditions. Any sudden cooling will condense steam and result in a pressure decrease, so simultaneous creation of high pressure and low temperature is improbable. ***Also contributing to the lack of PTS concerns for BWRs is the lower fast neutron fluence at the vessel inner wall, and the use of a thinner vessel wall which results in a lower stress intensity for a postulated crack.***"

The ***emphasized*** observations from SECY-82-465 hold true today: BWRs characteristically exhibit much lower embrittlement than PWRs as a direct consequence of the larger water gap between the core and the vessel wall (BWRs have approximately twice the diameter of PWRs), which reduces fluence and (thereby) irradiation damage to the RPV steel. Additionally, BWRs have thinner walls than PWRs, which reduces the magnitude of the thermal stresses. It may also be noted that our current findings reveal (see Section 8.3.5 of this document) that it is only at ***very*** high levels of embrittlement (many many 20-year license extensions) that thermal-only transients (large- and medium-break LOCAs) contribute significantly to the risk of driving a crack through the vessel wall.

Taken together, these observations suggest that failure of BWRs by thermal-only transients is highly unlikely.

Reply to Reviewer Comment #30

Comment made by: Johnson

Reply by: PRAB

Comment: I am quite interested in understanding how the information from the PRA portion of the analyses is "passed" to the thermal-hydraulic analyses. The "PRA Procedures and Uncertainty for PTS Analysis," draft letter report, October 2003, describes a binning process for the PRA results. It is not clear how these bins also formed the analysis boundary between the PRA and the thermal-hydraulic analyses. I will continue to explore this.

Staff Response: Sections 5.2.4 through 5.2.7 of this report provide a high-level description of the various activities associated with the binning process. Initial TH calculations (i.e., bins) were examined to determine which PTS scenarios would "match" the TH conditions of the TH bins. PTS scenarios that "matched" or were expected to be similar to the TH conditions were grouped into the existing TH bins. If the PTS scenarios were expected to have TH conditions that would be "different," or if the analysts (PRA and TH) were uncertain that conditions would be different, a new TH calculation was performed. If the TH conditions were different, then those PTS scenarios that were expected to have similar conditions were grouped into the new bin. This iterative process between PRA and TH continued until all potential PTS scenarios were allocated to the TH bins. Thus, there was no formal "passing" of information from the PRA to the TH; rather, information "passed" informally between (i.e., to and from) the PRA and the TH to create the set of TH bins for each analysis.

Reply to Reviewer Comment #31

Comment made by: van Walle

Reply by: MEB

Comment: General editorial comments:

A. Document (1: Dec. 2002 Draft NUREG) and (2: PFM Oct. 2003 Report) need a reference list of abbreviations or acronyms to facilitate readability;
B. Both documents are well written, but contain some typographic errors. Confusion can exist on the consistency of notations of symbols in text, formula and figures throughout the texts. Moreover a number of references are not well worked out, this also counts for some footnotes. Sometimes figures should follow their text closer.
C. Some overall revisiting should be done to avoid duplication of certain parts in the texts;
D. Although most of the flow diagrams in the text, especially in report (2), are impressive and help the reader to get the overall picture, some are quite confusing. This can be solved by more rationalization initially and further refinement as the text goes along. A flow diagram on FAVOR in text (2) would be useful too.

Staff Response: Editorial revisions have been made throughout the documentation to address these comments. Our nomenclature has been clarified by adding both Abbreviations and a List of Symbols to this document.

Reply to Reviewer Comment #32

Comment made by: van Walle

Reply by: MEB

Comment: Any possible 'relaxation' of the actual 10 CFR 50.61 rule to the proposed new ruling, as suggested in the initial part of the executive summary of document (1) with the words "without imposing on the licensees either new material testing requirements or new inspection programs", can not be made unconditionally for the NPP's. Within the actual PRA approach considerable importance is given to (a) operator experience and actions, that evolve from training activities; (b) more technical information resulting from inspections of components and bookkeeping of events that contribute substantially to justify a decreasing risk for PTS to occur. The reviewer's opinion is that when a NPP exceeds in future the actual 10 CFR 50.61 ruling, the NPP should implement a continuity plan - details to be discussed at a later moment - for training and inspections in order to being granted extra PTS 'margin' in accordance to the new ruling. This way the NPP can 'profit' from the new PTS Rule but shall maintain his 'good practice' or increase his efforts on training and inspection in order to guarantee future accordance with the principles that led to the 'relaxation' of the 10 CFR 50.61 PTS Rule;

Staff Response: The issues raised in this comment would be addressed by NRR as part of the rulemaking process, should rulemaking be undertaken.

Reply to Reviewer Comment #33

Comment made by: van Walle

Reply by: MEB

Comment: In general, the technical workout of documents (1) and (2) frequently contains 'soft' words or phrases such as 'are not expected to contribute to the risk', 'simply', 'approximate', 'unfortunately',... that are used to justify important reasoning. Wherever possible, these wordings should be omitted and quantified in numbers;

Staff Response: We have quantified our results where such quantification is possible.

Reply to Reviewer Comment #34

Comment made by: van Walle

Reply by: MEB

Comment: The technical workout of the PFM methodology contains a number of explanations of phenomena - like crack initiation, arrest, ... - that are aimed at providing a good, sometimes fundamental, physical understanding of what happens within the materials. Although the effort of the authors to provide clarity to the reader is highly appreciated and somehow needed, some of these explanations are still subject to discussion in the scientific community and should not appear as detailed in the texts. The reviewer will provide detailed comments to the authors.

Staff Response: We have revised the technical documentation in both this report as well as in [***EricksonKirk-PFM***] to achieve the dual goals of (1) providing a comprehensive technical description of our modeling approach and of (2) fairly reflecting the degree to which the models we have incorporated into FAVOR are generally accepted by the technical community.

Comment made by: van Walle

Reply by: MEB

Comment: Along the same line, it is not always clear how and up to which degree some input parameters and their uncertainties to the PFM routine are 'discretized' within the PFM routine FAVOR (difficult to find out in the FAVOR text or manual (3));

Staff Response: Again, we have attempted to improve the clarity of our writing, especially regarding the treatment of uncertainties and how they are modeled in FAVOR. The text in [***EricksonKirk-PFM***] and [***Williams***] have been revised accordingly.

Reply to Reviewer Comment #36

Comment made by: van Walle

Reply by: MEB

Comment: Some of the models and associated methodologies in the PFM part are rather easy to accept and are well established, others are still under discussion in committees, and some are rather new and are looked at with some skepticism in the scientific community. The main reason is that the authors and the NPP community has (and wants) to live with the measured information from the existing or executed surveillance programs within the NPP's. The reviewer understands this difficulty. However, the consequence is that correlation methods are a central issue in the PFM models. Unmistakably, correlation methods introduce uncertainty and an increased risk for error propagation. Moreover, correlation methods depend on statistics and fitting methodologies. Therefore, there should be a continuous effort to try to optimize the established trends and in a number of cases the collection of more data is advisable. This is especially true for fracture toughness data which form an important decisive data set in the PTS-rule evaluation.

Staff Response: We agree completely with the Dr. VanWalle's view that correlation methods introduce uncertainties, sometimes large uncertainties, into the computational models. We also agree that alternative procedures to directly measure the fracture toughness and embrittlement properties of the vessel materials (the Master Curve method, for example) show great promise in reducing these uncertainties. These methods should, and are, being pursued in other programs. However, for the reasons pointed out in Section 3.2.1 of this report, the use of such methodologies is beyond the scope of this particular project.

Comment made by: van Walle

Reply by: MEB

Comment: Do the three plants used for the PTS reevaluation project represent the entire fleet of US-power plants? Are there somewhere sequences out there in less embrittled plants, with drastically different operator actions (due to design for example), with different flaw distributions, with different limiting materials, with less info on surveillance data that are not represented by the four plants: some of these plants will also embrittle in their lifetime up to a level of the actual four plants. Does the rule envelop those situations?

Staff Response: While the three plants we have analyzed in detail contain some of the most embrittled materials in the operating PWR fleet, they are not the most embrittled (nor were they ever intended to be). As described in Chapter 9 (and especially in Section 9.3), we have examined a larger set of PWRs with the aim of identifying (at least qualitatively) the likelihood that the total population of PWRs contains situations where both more severe transients and more embrittled materials exist at the same plant. As described in Section 9.3, we believe that the information reported herein and in [***Whitehead-Gen***] provides reasonable assurance that the likelihood of both more severe transients and more severely embrittled materials occurring coincidentally at the same plant is remote. Consequently, we believe that the detailed results we have presented for Oconee Unit 1, Beaver Valley Unit 1, and Palisades provide an appropriate basis for establishing a PTS screening criteria that applies to PWRs *in general.*

The specific question of the general applicability of the flaw distribution we have assumed is addressed in detail in both Appendix C of this report, as well as in [***Simonen***]. As stated in those documents, we believe that the flaw distribution adopted by FAVOR is either an appropriate or a conservative representation of the flaw distribution that exists in any domestic PWR.

Reply to Reviewer Comment #38

Comment made by: van Walle

Reply by: MEB

Comment: The PTS reevaluation project splits uncertainty treatment in two kinds: aleatory and epistemic. This treatment is in principle very adequate, but the separation of the uncertainty kind in a variable is not always straightforward. How does one treat 'mixed' uncertainties and what are the consequences? Can we simulate this?

Staff Response: Currently, mathematical procedures do not exist to treat "mixed" uncertainties. In the process of model building, one must make the judgment that a particular uncertainty is "mostly" aleatory or "mostly" epistemic. In our reports detailing the technical basis for our models, we have described the technical bases for these judgments.

Reply to Reviewer Comment #39

Comment made by: van Walle

Reply by: MEB & SMSAB

Comment: All of this comment pertains to the crack initiation model

A. The PTS reevaluation project is based on probabilistic calculations. Yet some input values are treated deterministically (in principle when the uncertainty connected to them is epistemic): a major example that can influence the PTS-evaluation seriously is the output of the TH routine: pressure, temperature and heat transfer coefficient are so-called best estimates. It is said that the uncertainty on these best estimates is treated in the RELAP code and the binning. Does this mean that every best estimated value that comes out of RELAP (within a specific bin) has an uncertainty distribution that is by random selection propagated thru the FAVOR code?

B. WPS: in principle it is appropriate to include this effect in the PRA approach, but with the information given it is not easy to see how the justification is made for the deterministic entry in FAVOR. The uncertainty in KIc is aleatory, the mostly epistemic uncertainty in Kapplied seems not so difficult to characterize if the TH information is reliable and reliably transferred to the vessel wall. How is this treated? We may not forget that, apparently, the WPS introduction gives a major effect in the PTS Rule relaxation. Can this be quantified: what if WPS is not included, how sensitive is the whole procedure to the uncertainties in the parameters involved?

C. How reliable are the TH calculations to allow for a spatial distribution of its output parameters that can be reliably 'discretized' in FAVOR?

D. The theoretical basis for the universal temperature dependency of fracture toughness is rather well established these days (although some data demonstrate an apparent shape change at high irradiation level: still subject of discussion), at least good enough for the PTS risk. The reviewer would just advise to be less detailed as some of the physics is still open for discussion;

E. The statistical basis available to conclude that the scatter on fracture toughness KIa is smaller than the one on KIc is not well established, neither the theoretical explanation. The trend is there, but that's it (statistics?);

F. The reviewer agrees that T0 is the best estimate for initiation fracture toughness. This leads to the modification of the MC procedure for LEFM consistency. Although the procedure to quantify the epistemic uncertainty in itself seems justifiable (although not ideal), it is unsatisfactorily explained: textual confusion exists on the adjusted lower bounding curve (to all data of the 18 heats, to all data of one heat, who says 'coincidence' with the lowest KIc value is 'lowest'). The merit, but the loss of beauty given by the size effect adjustment, is compensated by the valid KIc values. A question can be raised towards the statistics (18 heats) used to define the modification; It is a pity that we cannot start from the real T0 values or the MC procedure. How sensitive is the whole procedure when we omit some heats, or would add some other material data?

G. The reviewer does not fully understand the use of the RT_{NDT} adjustment as being appropriate for the generic RT_{NDT} values: is the estimation of a bounding RT_{NDT} value from a generic one simply the random selection of an RT_{NDT} value from a normal distribution having the same mean and standard deviation as the data set originally used to establish the generic RT_{NDT} value or do we need to take a lower bound?

H. Within the index temperature shift model the uncertainty on the fluence attenuation thru the wall is not accounted for: discussion with neutron physicists leads to the suggestion that this uncertainty should at least be on the order of the precision by which neutron fluence at the RPV-wall can be determined: i.e., 10-20%, even when stated that the relationship is conservative;

I. The Charpy irradiation shift model: it is clear that the Eason trend shift model contains the general trends of most of the up-to-now recognized damage models. However, nowadays within the ASTM community there exists a lot of discussion on the appropriateness of the data sets that went into the model: does one need to separate BWR from PWR data? What about flux effects, long time ageing? The other question relates to the uncertainty treatment: in order to avoid 'double counting' no simulation of the uncertainty in the model is performed. Please remember that within the data sets used for fitting the model, other uncertainties than the ones on the chemistry content remain present: bias between testing and analysis techniques, temperature effects, different reactor type irradiation,.... Within ASTM and EPRI the discussion on which data sets to include and which to omit is still very much alive....Clearly some uncertainty needs to be accounted for;

J. Conversion of Charpy shift to Toughness shift: the physical grounds that state that the thermal-mechanical processing related to product form plays no role in temperature shifts would be true if we wouldn't have thermal ageing effects in materials at operational temperature. Even without irradiation the initial heat treatment of the product form that determines the microstructure of the material may alter when the material 'sits' for long time at operational temperature: so the statement in 3.2.3.4.1 is not fully correct;

K. What is the justification for not taking any uncertainty correlated to the conversion of Charpy-shift to toughness? Clearly the explanation given by the authors is not satisfactory: only statistics can reduce the standard deviation on these relationships. Omission does not seem to be justified within this type of correlation;

Staff Response:

A. The RELAP5 output of pressure, temperature, and heat transfer is always scenario-specific. FAVOR cannot accept these parameters as distributions. Nor would it make physical sense to do so, rather, the vessel temperature distribution must reflect the entire time-history of a transient. The heat flux and, therefore, temperature and heat transfer, can only be input as time histories for the calculation of temperature distribution to have any validity. Additionally, the time history of pressure is dependent on the time history of the entire scenario. Nonetheless, to elucidate the effects of variations in temperature and heat transfer, sensitivity studies were performed, as summarized in Section 9.1 of this report.

B. As detailed in Section 4.2.1 of [***EricksonKirk-SS***] and [Dickson 02], the overall result of including warm prestess effects in our model is a reduction in the TWCF of between a factor of 2 and 3, a significant but not major effect. Comparisons of RELAP predictions to experiments (see Section 6.7 of this report and also [***Fletcher***]) reveals that while RELAP predictions of pressure and temperature generally agree well with measurements, the differences between RELAP predictions and experiments that do exist are neither systematically high nor low. Consequently, any errors in the TH inputs to FAVOR will not cause the WPS model to systematically over or under estimate the failure probabilities.

C. See response to Comment #18.

D. Editorial revisions have been made to Section 3.2.1 of [***EricksonKirk-PFM***], as appropriate.

E. Editorial revisions have been made to Section 3.2.1.2.2 of [***EricksonKirk-PFM***] to more accurately describe the basis for the uncertainty in crack arrest toughness assumed by the model in FAVOR.

F. Editorial revisions have been made to Section 3.2.2.3 of [***EricksonKirk-PFM***] to better describe the procedure adopted by FAVOR to correct (at least approximately) for the bias in RT_{NDT}. With regard to the statistical adequacy of the empirical basis for this procedure, it should be noted that the data used to develop FAVOR's bias correction agrees well with a larger set of data compiled by European researchers [Houssin 01].

G. As detailed in Section 3.2.2.3.1.2.2 of [***EricksonKirk-PFM***] and in [***Williams***], Dr. VanWalle is correct in his statement that when a RT_{NDT} value is identified as "generic" FAVOR begins the simulation by selecting at random a RT_{NDT} value from a standard normal distribution that has as it's mean the generic RT_{NDT} value. This procedure simulates appropriately the uncertainty associated with generic RT_{NDT} values, essentially re-creating the data set from which the generic RT_{NDT} value was originally derived. Alternatively, one could select an upper bound RT_{NDT} value in all cases and arrive at a conservative representation, but such a methodology would be inconsistent with the "best-estimate" approach we have tried to adopt, to the greatest extent practicable, throughout this project.

H. Dr. VanWalle is correct in stating that uncertainty in our fluence attenuation model is not accounted for. However, for reasons stated in Section 3.2.3.1.3 of [***EricksonKirk-SS***] and in Comment 73 in this document, considerable evidence exists that the fluence attenuation model in Regulatory Guide 1.99 Revision 2, which we adopt in FAVOR, is a conservative representation reality. Indeed, this view is supported by a recently published survey of neutron attenuation models and data [English 02]. Because of the conservatism inherent in our fluence model, the staff does not view the treatment of uncertainties in the attenuation model as necessary. It should also be noted that FAVOR does account for the effects of uncertainty in our estimates of inner diameter fluence that are based on the procedures of Regulatory Guide 1.190. As detailed in Section 3.2.3.1.2 of [***EricksonKirk-SS***], the magnitude of fluence uncertainty used in FAVOR is consistent with that suggested by Dr. VanWalle.

I. In Section 9.2.1.2.3, we present the results of a sensitivity study wherein we have used the ASTM E900 embrittlement trend curve rather than the [Eason] embrittlement trend curve. This study demonstrates that the [Easton] model predicts TWCFs approximately 3 times larger than the ASTM E900 model. We have retained the [Eason] model as an implicit conservatism in our model, awaiting ASTM consensus on a new revision of E900 that is likely to be different than either Eason 00 or E900 is today. Additionally, Section 9.2.1.2.4 presents information confirming the appropriateness of FAVOR's approach to uncertainty simulation for the embrittlement model.

J. See explanation in Section 3.2.2.4.1 of [***EricksonKirk-PFM***] regarding the lack of product form dependency in fracture toughness data.

K. See explanation in Section 3.2.2.4.1 of [***EricksonKirk-PFM***] regarding the origin of uncertainties in the Charpy → toughness conversion, and why it is appropriate to view this relationship as being, for all practical purposes, without error.

Comment made by: van Walle

Reply by: MEB

Comment:

All of this comment pertains to the upper shelf ductile tearing model

A. Ductile tearing definitely is an option to occur on existing flaws or after crack arrest of a running crack. In principle, the information for the modeling of this effect should directly come from experiment. In view of the non-availability of this information a model based on USE properties is constructed. The reviewer is quite reluctant to accept this model as its partly empirical correlations, certainly not all accepted by the scientific community, have large uncertainties that are not taken into account. The functionalities for the correlations demonstrate trends, but the data are widespread;
B. The M1 model based on the Eason relationship is open for a lot of discussion. The reviewer does not really believe this model. Maybe it is the best option if one has to take it (better than RG1.99), but the fact that for example the initial heat treatment of the material doesn't have an influence on the decay of the upper shelf seems strange. The conclusion on the as-defined USE that 'the data demonstrate that the spread in individual USE values scales in proportion with the absolute magnitude of the USE, and that the uncertainty in upper shelf energy is essentially unaffected by irradiation' may be remarkable, but please remark that the USE spread at every 'mean' upper shelf energy in the data set is about half the absolute value of the upper shelf. Also remark that the statistics on the data at the 150 ft-lbs USE level, that determines the trend, is extremely low
C. The conclusion on the uncertainty treatment for the probabilistic upper shelf tearing model on best estimate relationships is simply unacceptable if it only takes the variation in chemistry and fluence into account. The reviewer understands that the uncertainty levels go way up if the spreads on all correlations are taken into account, but the consequences should be looked at in a probabilistic approach;
D. Material property gradient model: the reviewer understands the reasoning of the influences of K_{Ic} and K_{Ia} within the initiation, arrest, re-initiation approach and the methodologies used for bounding seem acceptable. The problem for the reviewer lies again in an insight on the sensitivity of the whole PTS procedure to variations in these approaches;
E. On the gradient composition within welds, the reviewer would just like to see what the influence on the PTS procedure is if you don't take it into account: isn't it an effect of second order?

Staff Response:

In comments (A) – (C), Dr. VanWalle questions the adequacy of the staff's model of ductile fracture toughness on the upper shelf, which was based in large part on correlation with upper shelf Charpy V-notch properties. After review of these comments and further discussions with Dr. VanWalle, the staff adopted a new upper shelf model and implemented it in FAVOR to address these comments. This new model does not rely on Charpy correlations in any way, and features an explicit treatment of the uncertainty in upper shelf toughness (both the ductile initiation toughness as measured by J_{Ic} and the resistance to further crack extension as measured by *J-R*). This upper shelf model is based on work recently completed by EPRI [EricksonKirk 04]. Details of the FAVOR implementation of this new model can be found in Section 4.2 of [***EricksonKirk-PFM***] and in [***Williams***]. The effect of adopting this physically based upper shelf model rather than the correlative model used previously is discussed in Section 9.2.1.2.7.

Regarding Comment (D), alternative models that assume no linkage between K_{Ic} and K_{Ia} would allow an initiated crack to immediately arrest because a K_{Ia} value could be simulated that exceeds the K_{Ic} value that produced the crack initiation. As discussed in [***EricksonKirk-PFM***], we do not feel that this alternative model is appropriate on physical grounds. Additionally, it can be noted that the model adopted in FAVOR is conservative relative to the alternative model that assumes no linkage between K_{Ic} and K_{Ia}. For this reason, we have not performed a sensitivity study to address this question.

Regarding comment (E), we have performed a sensitivity study on the effect of the through-wall composition gradient in welds. The results of this analysis (see 9.2.1.2.5) show that if through-wall composition gradient is removed from the FAVOR model the TWCF increases by a factor of approximately 2½.

Reply to Reviewer Comment #41

Comment made by: van Walle

Reply by: MEB

Comment: Conclusions: The reviewer recognizes the important effort that has been put into the PTS reevaluation method and believes it to be a necessary step to come to a more realistic basis for plant life assessment. The comments on the proposed methodology, concentrated on the PFM part, can be summarized as:

A. The general methodology for the PTS reevaluation process can be accepted;
B. The models that go into the PFM-study are in part subject for discussion, as treated above. As a number of models are based on correlations, statistics and uncertainty evaluation remain important issues;
C. Overall the uncertainty treatment needs to be more worked out and justified;
D. Sensitivity studies can be an important asset to single out the parameters of importance.

Staff Response: These are general comments that have been addressed previously.

Comment made by: Rohatgi

Reply by: SMSAB

Comment: A PIRT was developed. However, many of the items listed are not phenomena but boundary conditions or operator actions. These do have influence on the downcomer conditions and should be considered.

Staff Response: It is certainly true that many items in the PTS PIRT are boundary conditions. This is true of all thermal-hydraulic analyses. We considered both phenomena and boundary conditions in our uncertainty evaluation.

Comment: RELAP5 is in general applicable to predict reactor system behavior. I will have difficulty in accepting multidimensional capability, in downcomer and in cold leg. That deficiency in the code has to be treated as bias. Also, the heat transfer coefficient prediction in the downcomer may not be appropriate.

Staff Response: As a one-dimensional code, RELAP5 does not provide true multidimensional calculational capability. However, applicable experimental data were reviewed from integral system test facilities LOFT, ROSA, APEX-CE, as well as full-scale tests from UPTF and reduced-scale separate effects tests from Creare, IVO, and Purdue University. The data show that the RELAP5 modeling of the downcomer is reasonable. The integral system experimental data were used to determine biases and standard deviations for the RELAP5 calculation of downcomer temperature, downcomer heat transfer, and system pressure. The experimental data show that large temperature gradients in the cold leg (i.e., ~ 100°C) do not translate into corresponding plumes in the downcomer, because of large eddy mixing occurring in the downcomer volume. Introducing a bias is only appropriate if the code is shown to be nonconservative, which was not the case for all three parameters of pressure, temperature, and heat transfer. This is discussed further in [***Bessette***].

Comment made by: Rohatgi

Reply by: SMSAB

Comment: The NUREG report for RELAP5 assessment has a PIRT description. However, there is no correspondence between high-ranking phenomena and tests modeled with RELAP5. Also, the assessment is qualitative. How are the assessment results factored in the uncertainty analyses?

Staff Response: There is, in fact, consistency between the PIRT and the assessment and uncertainty analyses that were performed. Description has been included in Chapter 6 of this document to show the correspondence between phenomena, assessment, and uncertainty analyses. Additional description is included in [***Bessette***]. The integral system tests assessment results were evaluated statistically to generate means and standard deviations showing the predictive capability of RELAP5 for downcomer temperature and system pressure. Uncertainty analyses and sensitivity studies were performed to evaluate the importance of key phenomena, including break flow, heat transfer modeling, downcomer temperature, accumulator modeling, natural circulation flow, downcomer-upper plenum bypass, and steam generator heat transfer. Separate effects assessment was performed for important phenomena identified by the PIRT. This is discussed in NUREG-1809.

Reply to Reviewer Comment #44

Comment made by: Rohatgi

Reply by: SMSAB

Comment: There is need for a discussion, why a temperature and heat coefficient distribution (or local values) are not important for CDF?

Staff Report: This is discussed in [***Bessette***]. The simplifying approximation of using a uniform distribution for temperature and heat transfer is shown to be reasonable and sufficient.

Comment made by: Rohatgi

Reply by: SMSAB

Comment: RELAP5 uses (66) nodes to simulate flow in the downcomer. How accurate is RELAP5 in predicting TH conditions at different locations. Has any comparison been made with CFD or tests? UPTF tests were performed in support of LB LOCA and there were temperature measurements in downcomer. This data could be of use.

Staff Response: Applicable experimental data were reviewed from integral system test facilities LOFT, ROSA, APEX-CE, as well as full-scale tests from UPTF and reduced-scale separate effects tests from Creare, IVO, and Purdue University. The data show that a two-dimensional modeling used in the RELAP5 PTS calculations is reasonable. The non-uniform temperature distributions in the experimental data are within the absolute uncertainty of RELAP5 to predict bulk fluid temperature. In fact, the temperature variations seen in APEX-CE experiment are of the same order (5°C) as RELAP5 predictions of plant transients.

Comment made by: Rohatgi

Reply by: SMSAB

Comment: How much variation is expected in the TH conditions in the downcomer? Is this variation captured through uncertainty analyses?

Staff Response: Temperature variations in the downcomer calculated by RELAP5 are on the order of 5°C; see Chapter 5 of [***Bessette***]. This is also seen in the relevant experimental data. The variation is far less than that considered by the treatment of uncertainties.

Comment made by: Rohatgi

Reply by: SMSAB

Comment: It will be interesting to see OSU tests as they were used in making some judgment on PIRT. OSU tests showed that mixing in cold leg increased the fluid temperature enough to reduce the severity of the transients. How valid is this mixing in the cold leg? RELAP5 cannot predict this. Do we account for this in code bias?

Staff Report: The APEX-CE tests showed temperature gradients across the cold leg of up to 120°C (200°F). The cold leg diameter in the facility is 9-cm (3.5-in.). With this thermal stratification in the cold leg (200°F on the bottom and 400°F on the top of the cold leg), the temperature nonuniformity in the downcomer, both axially and azimuthally, was on the order of 5°C. Comparison of RELAP5 with integral system test data for downcomer temperature shows no code bias (RELAP5 ~3°C conservative).

Comment made by: Rohatgi

Reply by: SMSAB

Comment: There were calculations done for ROSA facility with RELAP5 and there is large temperature variation (around 200C) in the cold leg (Fig. 3-24, 3-25, 3-33). How is this reflected in the downcomer temperature and heat transfer coefficient? RELAP5 underpredicted cold leg temperature and overpredicted downcomer temperature. Is there an explanation?

Staff Response: RELAP5 cannot calculate temperature gradients in the cold leg since it only represents and single liquid field in any given node. In this situation, RELAP5 tends to predict an averaged behavior. The RELAP5 calculations of cold leg temperatures in ROSA generally fall within the temperature distribution measured in the cold leg. As consistently seen in the experimental data, the downcomer is well-mixed. The large temperature gradients in the cold leg do not translate to corresponding temperature variations in the downcomer. This is attributable to large eddy mixing occurring in the downcomer. As noted, RELAP5 comparisons with ROSA experiments for downcomer temperature show excellent accuracy (RELAP5 ~4°C conservative).

Comment made by: Rohatgi

Reply by: SMSAB

Comment: ROSA tests also had flow reversal in cold legs (3-5, 3-6) that RELAP5 did not predict?

Staff Response: There is no physical reason for reverse flow. The experimental data are from venturi flow meters located in the cold leg loop seals. The RELAP5 calculations are believed to be accurate. The flow measurements are believed to be inaccurate under two-phase, low-flow conditions, where there may be a zero-shift. In practice, there is back flow from the ECC injection locations to the loop seal region, where there is fluid-fluid mixing. However, there is no cold leg to cold leg loop flow between pairs of parallel cold legs attached to a single steam generator.

Comment made by: Rohatgi

Reply by: SMSAB

Comment: RELAP5 overpredicted semiscale flows. That could due to smaller loss coefficients or higher interfacial mass transfer. Is this effect factored in uncertainty analyses?

Staff Response: The purpose of the particular SEMISCALE experiments was to evaluate natural circulation flows under reduced primary system inventory conditions. The experiment was run without a break in the system, with the inventory reduced in steps. The overprediction of mass flow rate in the RELAP5 calculation of two-phase natural circulation is a result of overprediction of interphase drag under bubbly flow conditions. Two-phase natural circulation was included in the set of assessment cases because it was included in the original PTS PIRT, but is not particularly significant, since this heat transfer regime exists only for short times, and has essentially no impact on system pressure, downcomer temperature, and downcomer heat transfer.

Comment made by: Rohatgi

Reply by: SMSAB

Comment: Heat transfer coefficient is based on Dittus Boelter and that is for pipes for fully developed flow. Downcomer flow is different and may have counter current flow due to mixed-convection. How much is the uncertainty in applying Dittus Boelter correlation?

Staff Response: For the conditions of interest to PTS, RELAP5 applies Churchill-Chu free convection heat transfer modeling and Dittus-Boelter for higher velocity forced convection conditions (it applies the maximum of the two models). The flow velocities seen in experiments in UPTF, APEX, and Creare are substantially higher (0.3 to1.5 m/s) than might be applicable to mixed-convection conditions. Therefore, a low-flow mixed-convection condition does not apply to the downcomer [***Bessette***].

Comment made by: Rohatgi

Reply by: SMSAB

Comment: It is not clear how average downcomer temperature (averaged over space and 10,000s time period) will provide the uncertainty in the T_{dc} prediction? How is this information used?

Staff Response: Sensitivity studies were run with RELAP5 utilizing the PTS PIRT. The studies varied both boundary conditions and physical models. Each parameter was varied from its nominal, or best estimate value, to and upper and lower value, based on the uncertainty of the particular parameter. These sensitivity studies were performed one parameter at a time. The downcomer temperature was chosen as the figure of merit for these calculations. Temperature is the most important factor influencing the conditional probability of vessel failure, as compared to pressure and heat transfer coefficient, ergo its selection. The downcomer temperature was averaged over the 10,000s time duration of the calculation. The importance of each parameter was measured by its affect on the time-averaged value of downcomer temperature, over the 10,000s of the calculation. This choice of 10,000s as the time-averaging interval is subjective, but is reasonable given the timing of vessel failure for the spectrum of PTS sequences analyzed. Some sequences had peak failure probabilities as early as 350s, while others had peak failure probabilities occurring at approximately 7500s.

The time-averaged downcomer temperature was used as the importance measure for each parameter that was evaluated through the sensitivity studies. From this, a computer code was used to combine all the parameters using Monte Carlo methods to generate an uncertainty distribution for each PRA bin of PTS sequences. From this distribution of temperatures, discrete event sequences were selected that, as calculated by RELAP5, represent the range of outcomes that characterized each bin. The total probability of sequences that fell within the bin, as defined by PRA, was subdivided and apportioned. The multiple RELAP5 runs were used to characterize the bin.

Further description of the basis for the 10,000s interval is given in Section 6.9 of this report and in [***Chang***].

Comment made by: Rohatgi

Reply by: SMSAB

Comment: The conduction model in RELAP5 affects the temperature and heat transfer coefficients. PFM also performs conduction calculations using these as boundary conditions. Are there any differences in conduction models in two codes and how they affect the temperature/ stress prediction in the wall?

Staff Response: The RELAP5 conduction equation has a feedback effect on the determination of the convective heat transfer coefficient. This was evaluated by performing vessel wall nodalization sensitivity studies. FAVOR solves the conduction equation itself and does not utilize the RELAP5 conduction modeling. FAVOR has been benchmarked against ABAQUS, with excellent agreement [Dickson 03b].

Comment made by: Rohatgi

Reply by: SMSAB

Comment: RELAP5 assessment reports indicated a PIRT. However, it will be useful if the report could show that the assessment tests were connected to high-ranking phenomena

Staff Response: This information is addressed in Section 6.7 of this report.

Comment: Ranging of parameters for high-ranking phenomena, were done but it is not clear how assessments contributed to the ranges.

Staff Response: The phenomena evaluated as part of the uncertainty analysis were convective heat transfer, accumulator injection, break flow, and natural circulation. Convective heat transfer was varied by 30%. The Henry-Fauske break flow modeling was ranged by 30%, which is a generally accepted value. The accumulator injection was varied by varying the pressure 50 psi, to encompass the uncertainty in pressure and flow modeling. Natural circulation was varied by change the loop resistance by factors of 0.5 and 2. Additional sensitivity studies were performed with respect to vessel wall heat conduction, convection heat transfer, and upper plenum/downcomer bypass flow. Additional assessment and analysis were performed with respect to steam generator heat transfer, condensation, and flow distribution and mixing in the cold legs and downcomer.

Comment made by: Catton

Reply by: MEB

Comment: It would have been helpful to have seen a summary that highlights the uncertainties in the input to the thermal-hydraulics, the uncertainties in the input to the structural mechanics and the final failure probability and uncertainty resulting from the structural mechanics output for a few key transients.

Staff Response: Section 3.2 of this report was added to address this question. This section describes how uncertainties are addressed in PRA, TH, and PFM analyses and how these unçertainties are "propagated" through the analyses.

Reply to Reviewer Comment #56

Comment made by: Catton

Reply by: SMSAB

Comment: Arguments were given as to why the LOCA is more important than the events previously thought to dominate initiators. The basis for this surprising outcome is the role of the "subcool meter". Before I become a proponent of this view, I will need to learn more about "subcool meters". Here I only need to remind you of TMI and the temperature measurements available in the control room during the accident.

Staff Response: The reasons for the importance of LOCAs versus secondary side failures include the following: (1) LOCAs produce severe cooldowns that have relatively high conditional probabilities of failure (CPFs); (2) operators cannot mitigate cooldown of medium and large LOCAs (coolant injection is required to prevent core damage); and (3) operators have little time to respond to stuck-open safety relief valves that reclose. A subcooling meter reduces the importance (i.e., the frequency) of scenarios involving secondary side failures by providing operators with information that can be used to distinguish between and respond to the events, thus reducing human failure probabilities. Also, current fracture mechanics calculations indicate that secondary side failures are less important. The role and use of subcooling meters was also discussed at the peer review meeting on May 10–12, 2004.

Comment made by: Catton

Reply by: SMSAB

Comment: Both the wall flux and the time rate of change of the interface temperature are strong functions of the time history and magnitude of the heat transfer coefficient and fluid temperature. To show whether or not the heat transfer coefficient is important or not is a simple exercise whose solution can be found in any good book on conduction (e.g., Carslaw and Jaeger). Studies in the earlier visit to the PTS issue showed that the values of the heat transfer coefficient calculated using correlations now in the codes like RELAP5 fell midway between the wall conduction limit (very high heat transfer coefficient) and the convective limit (low heat transfer coefficient). The relationship between the heat transfer coefficient and failure probability for the base case was very steep.

Staff Response: This comment is addressed in Sections 6.8 and 9.1 of this report, and in [***Bessette***]. Additional analyses have been performed showing the role of convective heat transfer coefficient and the sensitivity of vessel failure predictions to heat transfer modeling and uncertainty.

Comment made by: Catton

Reply by: SMSAB

Comment: The twenty two phenomena were reduced to seven for consideration in determining the bounds or uncertainty in the thermal-hydraulic analysis. The heat transfer coefficient was ranked tenth leaving it out of further consideration. It was then argued that previous work and RELAP5/MOD3 development assessments had shown it to be adequate to predict these phenomena. For overall behavior, this may be true. Unfortunately, the downcomer has not received much attention in the past. Many facilities used a small pipe to represent the downcomer and for others it was a very thin annulus. Further, with the focus being on the core, the downcomer was seldom well enough instrumented to yield much meaningful data.

Staff Response: See response to Comment #57.

Comment made by: Catton

Reply by: SMSAB

Comment:

RELAP5/MOD3 with cross flow junctions in the downcomer is described and noted as an improvement (over what I do not know) for computation of flows in the downcomer. The use of cross flow junctions to represent multidimensional flow has long been known to yield erroneous results unless the cross flow is essentially zero. This issue was addressed more than twenty years ago by the ACRS. The COMMIX code was the CFD code of choice for more detailed studies. The amount of numerical mixing generated by the COMMIX code is difficult to quantify and it is not known if this was done. Another code mentioned was REMIX. REMIX is a simple multi-stage mixing code and is a reasonable tool but needs some comparison with experimental data for corroboration. The mixing parameters, and their uncertainties, used in REMIX can be estimated. These can in turn be used to estimate the uncertainty in temperature at various locations in the downcomer. This is promised in the PIRT report.

Staff Response:

The base case PTS calculations were performed with a multi-channel downcomer. This modeling has been found to provide a better representation of important phenomena. Specifically, the two-dimensional representation of the downcomer allows a degree of freedom not possible with a one-dimensional model. This provide a more stable calculation of flows in the different cold legs, and permits flow recirculation in the downcomer, with some degree of similitude to that observed in integral system experiments.

Sensitivity calculations were performed for Palisades comparing the base case two-dimensional downcomer with a one-dimensional downcomer. The comparisons between the one-dimensional and two-dimensional downcomer are for the most part very similar for hot leg breaks. For cold leg breaks, however, the one-dimensional downcomer model had warmer temperatures, since the nodalization forced more cold ECC injection to be bypassed out the broken cold leg. This tendency was nonconservative.

CFD codes in use to day have better turbulence modeling capability and less numerical mixing than the COMMIX code of 15 years ago.

REMIX has been extensively assessed against the available separate effects experimental data base for fluid-fluid mixing, including UPTF, HDR, Creare, Purdue, and IVO [NUREG/CR-5677]. The code is shown to be conservative in its predictions of plume strength in the downcomer. REMIX was used to calculate flow stagnation mixing and temperature distributions for Palisades [***Bessette***]. REMIX models the downcomer as the decay of a free plume in an otherwise well-mixed downcomer. It does not represent the large eddy circulation evident in the integral systems experiments. Therefore, it is conservative with respect to the downcomer mixing processes.

Together, the separate effects and integral system experiments and code calculations present a consistent picture. The available information shows that the uniform temperature distribution assumption is a valid simplifying approximation. This is discussed further in [***Bessette***].

Comment: What should be done to obtain an estimate of the heat transfer coefficient and its uncertainty is another matter. Because the process is mixed-convection, the resulting heat transfer coefficient depends on both the wall temperature and the fluid temperature.

Staff Response: See response to Comment #57. Mixed convection is not relevant to downcomer conditions during PTS scenarios.

Comment made by: Catton

Reply by: SMSAB

Comment: RELAP5 is a one-dimensional code developed with focus on LOCA events and the core of the reactor. In this case the vessel annulus is the dominant component of the event and the thermal-hydraulics in the annulus has never been the subject of much scrutiny. There have been surprises every time data becomes available. The French search for vibration sources in the thermal shield area led to the realization that there significant recirculation eddies exist under normal operating conditions. These are not important when heat transfer is not an issue. When data became available from the German contribution to the 2D/3D program, the annulus again produced surprises.

Staff Response: The assessment carried out in support of the PTS evaluation was focused on downcomer behavior, specifically temperature distribution. This included 2D/3D UPTF data on fluid-fluid mixing and condensation. Evaluation of integral system experimental data show extensive large eddy mixing processes (significant recirculation eddies) in the downcomer, indicating that the downcomer is will mixed. The result is that the temperature nonuniformity in the downcomer is on the order of 5°C. RELAP5 calculations of PTS scenarios provide similar (5°C) temperature variation in the downcomer. Downcomer velocities calculated by RELAP5 are consistent with the experimental data.

Together, the separate effects and integral system experiments and code calculations present a consistent picture. The available information shows that the uniform temperature distribution assumption is a valid simplifying approximation. The results show that there are no plumes of any practical significance. Temperature nonuniformity in the downcomer is small (on the order of 5°C) and is similar between RELAP5 and the experimental data. Furthermore, this temperature nonuniformity is within the uncertainty (10°C) of RELAP5 to predict averaged downcomer temperature. Therefore, further consideration of nonuniform temperature effects is not technically warranted. This is discussed further in [***Bessette***].

Comment made by: Catton

Reply by: SMSAB

Comment: When one uses a code like RELAP5, one needs to be suspicious and unsure of the results and a self avowed systems engineer cannot do this. There were others like Rex Shumway who supported the work but it is not clear to what extent. I was sent the section of NUREG/CR-5535-V4 that deals with heat transfer coefficient calculation as a follow up to earlier questions. In the report, mixed-convection is mentioned on pages 4-70 and 4-71 and then never seen again. For most calculations, RELAP5/MOD3 simply uses the Dittus-Boelter correlation. There is some discussion of the expected uncertainties in the heat transfer coefficient. It is noted that the correlation is ±25% for some cases. The correlation was developed for pipe flow and needs to be corrected when used for other geometries. There are adjustments noted for use in rod bundles but not for an annulus. Without considering mixed-convection, a minimum uncertainty would be 25%. This is larger than was used.

Staff Response: For the conditions of interest to PTS, RELAP5 applies the maximum of Churchill-Chu free convection and Dittus-Boelter for more turbulent forced convection flows. This is discussed in Section 9.1, in [***Bessette***], and in our response to Comment #6. Additional analyses have been performed showing the role of convective heat transfer coefficient and the sensitivity of vessel failure predictions to heat transfer modeling and uncertainty.

Comment made by: Catton

Reply by: SMSAB

Comment: RELAP5 gives an average temperature of some type, not a real value with the correct spatial distribution. We were told that the oscillations in temperature are important yet the RELAP5 results are given in 15 second intervals. It was not clear what was done with the intermediate points when the Probabilistic Fracture Mechanics (PFM) is done. When asked if this is conservative or not we were essentially told that it was sometimes one way and sometimes another. I frankly do not know what to do with information like this. Further, PFM is supplied with average pressure, temperature and heat transfer calculations evaluated at 15 sec intervals. The averages are over the entire internal vessel wall. Again, there is no way to decide whether this is conservative or pessimistic. It certainly is not best estimate.

Staff Response: Sensitivity studies were performed comparing the reporting frequency of 30s with 1s. Additional sensitivity studies were performed with an edit frequency of 1s, whereby the RELAP5 output was averaged over time intervals of 1s, 5s, and 10s. Negligible effects of edit frequency and of averaging were seen among the sensitivity cases; additionally, the results indicated neither systematic biases, either conservative or nonconservative. The results of these sensitivity studies are reported in Section 9.1 of this report and in [***Bessette***].

Comment made by: Catton

Reply by: SMSAB

Comment: The question becomes what to use from RELAP5 and how. One could plot a smooth curve above and below the RELAP5 predicted oscillating temperature and use both in a PFM analysis and see what the result is as a function of the frequency of oscillation between the two. If the results are not much different, then this is not a problem. This has been evaluated in part. The argument is that the h is well above 2500 and so the process is conduction limited. There are however circumstances where the pumps are off and natural circulation flows are low and the value of the heat transfer coefficient drops to 200 to 300. Many of the graphs in the material supplied shows values in this lower range. In this region, mixed-convection will become very important. It would be helpful to know what cases led to low h and where mixed-convection could be important..

Staff Response: This is discussed in Section 9.1 and in [***Bessette***]. Additional analyses have been performed showing the role of convective heat transfer coefficient and the sensitivity of vessel failure predictions to heat transfer modeling and uncertainty.

Reply to Reviewer Comment #64

Comment made by: Catton

Reply by: SMSAB

Comment: The uncertainty analysis did not include h as a parameter yet they included heat capacity. This is very unsettling and maybe that I missed something. I will review the appropriate reports and comment again if necessary.

Staff Response: See response to Comment #63.

Comment made by: Catton

Reply by: SMSAB

Comment: The basis for the experimental data chosen to validate or tune the code needs further amplification and justification. Need to be sure some of the tests have measured velocities in the downcomer, or at least enough data to back it out. No wall temperatures were measured in any of the past experiments with the exception of UPTF. This effort should have led to delineation of what was needed to evaluate the thermal-hydraulic uncertainties needed to evaluate uncertainties for a particular transient.

Staff Response: The rationale for the RELAP5 assessment is described in Section 6.7 and in [***Bessette***]. RELAP5 assessment included comparison with experimental data from UPTF and APEX. In addition, data from Creare were reviewed, where velocities were measured in the downcomer. RELAP5 has been compared with CFD calculations to compare velocity predictions. The data show a substantial enhancement (factor of 20) in downcomer superficial flows. An integrated assessment of RELAP5 predictions of downcomer heat transfer coefficient show them to be consistently conservative or realistic. No nonconservatisms were found in RELAP's estimates of heat transfer coefficient [***Bessette***].

Comment made by: Catton

Reply by: SMSAB

Comment: When bin uncertainties are discussed, the uncertainties are the result of slightly different cases being put in the same bin. This has nothing to do with code uncertainty. Within one of these bins, the code uncertainty should be shown to complete the argument. If the code uncertainty is well within the bin uncertainty, then the code is good enough. This has yet to be done.

Staff Response: Code uncertainty has been shown to be much less than "bin uncertainty." This is described in Sections 6.8 and 9.1, and in [***Bessette***]. In fact, a given bin includes a broad range of transients, as expressed in terms of downcomer temperature and system pressure. As shown in the main report, it is the variation within a bin that dominates the characterization of thermal-hydraulic uncertainty. This follows from the fact that thermal-hydraulic analysis is a function of boundary conditions as well as processes occurring within the control volume.

Reply to Reviewer Comment #67

Comment made by: Catton

Reply by: SMSAB

Comment: When the analysis was done it was assumed that there were no synergistic effects. One uncertainty at a time was assumed and then its impact was calculated. Only the impact on temperature was evaluated. This could be done because the inner vessel wall is around 3% of the total area. The question of how well the temperature in the downcomer can be calculated is yet to be evaluated.

Staff Response: The validity of RELAP5 predictions of downcomer temperature was established by comparing the code with a variety of integral system experimental data. The code was shown to predict downcomer temperatures with excellent accuracy (i.e., within 1%). Details are found in [***Fletcher***].

Comment made by: Catton

Reply by: SMSAB

Comment: Much of what was done is unsettling. For example, putting in a check valve between computational volumes in RELAP5 to stop what appears to be improper recirculation seems to beg the issue. If the code cannot get the flow direction right, how can one have faith in the temperature predictions.

Staff Response: A check valve was not used in the cold leg. Rather, the loss coefficient for flow in the reverse direction was introduced in the Oconee and Palisades models. The issue involves *circulation through identical, parallel pipes* in a piping network. For the current PTS analysis, the relevance of this problem is limited to the modeling of the cold leg regions of plants with two hot legs and four cold legs, namely plants of Babcock and Wilcox (B&W) and Combustion Engineering (CE) designs. In these plants, the cold legs are identical, in modeling terms. The concern arises in the absence of significant physical driving potential (i.e., pressure) for flow.

The cold leg flow circulation problem was identified initially during the Oconee analysis of the original PTS study (NUREG/CR-3761, pp. 88–90). The conclusion at the time (1984) was that, once initiated, this cold leg circulating flow was physical. Temperature differences in the cold leg fluid can, under certain conditions, provide a physical driving head to begin circulating flow. The physical mechanism begins with the injection of cold water into both cold legs between the reactor coolant pumps and reactor vessel. Mixing of the cold leg fluids can occur both in the reactor vessel downcomer and in the steam generator outlet plenum that is shared by the two cold legs. Because the cold legs are liquid filled, an incipient asymmetry can cause an imbalance in otherwise equal cold legs. The force balance may grow until flow reverses in one cold leg, with the adjacent cold leg then carrying both flows. In this situation, the fluid temperatures in the vertical sections of the cold legs are different, creating a buoyancy driving head that can sustain the circulating flow. Investigation into the numerical initiator indicated that flow first begins as a result of round-off errors (in the last digitally-stored significant digits) in the pressure solutions at the ends of the identical pipes. *The numerical initiator is, therefore, judged to be unavoidable when using a digital thermal-hydraulic systems code.*

The additional damping was introduced to prevent undue circulation. If circulation were present in a calculation, the effect would be to introduce an additional mixing process that, in this case, is more numerically induced rather than physical. This causes higher downcomer temperatures. Since higher temperatures may be nonconservative, the damping employed eliminated a potential nonconservatism. In fact, comparisons with data show the introduction of reverse damping produces a conservatism since it precludes the loop seals as a mixing volume, when in fact the loop seals participate in the mixing process.

It should be recalled that the accuracy of RELAP5 to predict downcomer temperatures was established through comparison with experimental data.

Reply to Reviewer Comment #69

Comment made by: Catton

Reply by: SMSAB

Comment: We heard arguments about binning, event uncertainties and how this made the thermal-hydraulic uncertainties insignificant. It is not a good policy to mix the uncertainties from one type of analysis with another. Uncertainties arising from the PRA become uncertainties in the boundary and initial conditions used in RELAP5.

Staff Response: A better justification to our treatment of uncertainties in both the PRA and TH areas appears in Section 3.2 and 6.9 of this report.

Comment made by: Catton

Reply by: SMSAB

Comment: There is a missing part to the story we heard. The PIRT should have led to statements about processes that are important. Comparisons of code predictions with data from facilities that have been shown (by a scaling analysis) to be relevant then lead to knowledge of the uncertainties in the predictions. The uncertainty in a code prediction needs to be evaluated by itself. Application to a plant with appropriate consideration of operational uncertainties then yields the value and uncertainty of the final result.

Staff Response: This is discussed in [***Bessette***].

Comment made by: Catton

Reply by: SMSAB

Comment: As a final comment, mixed-convection was not considered. As pointed out in my earlier note on this aspect of the PTS analysis, mixed-convection is very important and relatively low values of the natural convection effect can increase the heat transfer coefficient significantly. This could lead to a systematic underestimate of the potential for PTS.

Staff Response: This is discussed in Sections 6.8 and 9.1, and in [***Bessette***]. Mixed-convection is not relevant to conditions in the downcomer (see response to Comment #65).

Comment made by: Catton

Reply by: SMSAB

Comment: The advances in PFM since PTS was last addressed appear to be significant. When PTS was last addressed, the major uncertainties were in the PFM. It appears as if this has changed and that the values and uncertainties can be evaluated subject only to the uncertainties in the thermal-hydraulics. It came as some surprise, however, that the LOCA is a dominant contributor to the PTS risk.

Staff Response: We agree that substantial progress has been made in PFM since the previous study. This earlier study did not consider large-break LOCAs, which are an important contributor.

Comment made by: Schultz

Reply by: MEB

Comment: During the May 10-11, 2004 meeting the Peer Review panel Dr. Schultz commented that he felt the staff's proposed use of the Regulatory Guide 1.99 Revision 3 fluence attenuation function might be nonconservative relative to data of which he was aware. After the meeting the staff e-mailed Dr. Schultz to gain more information about and, hopefully, resolve the comment. In response Dr. Schultz's colleague, Dr. Uwe Jendrich, sent the staff a number of references on through-wall property measurements. Upon reviewing the references, the staff sent Dr. Jendrich the following response:

June 16th, 2004

Dr. Jendrich -

After reading over the references you sent it seems that this may all be a big misunderstanding (of which I am probably the source, so I apologize for that). It had been my understanding that Dr. Schultz expressed concern with our use of the Reg Guide 1.99 Rev. 2 attenuation function in our PTS reevaluation project because - and I ***THOUGHT*** this was what Dr. Schultz said - he claimed that data exist that show the RG attenuation function to be nonconservative (i.e., it OVER-predict the amount by which fluence, and therefore Charpy shift, attenuates with depth into the RPV wall). However, Fig 30 from your reference KUS-90 (on Gundremmigen) clearly shows the opposite to be the case: the RG prediction is CONSERVATIVE relative to the data because it UNDER-predicts the amount by which Charpy shift attenuates with depth into the RPV wall. Moreover, a report prepared recently by Colin English for [English 02] EPRI reaches the same conclusion (that if anything the RG attenuation function provides CONSERVATIVE (i.e., low) estimates of Charpy attenuation) based on a wider body of experimental evidence (see especially the figures on page A-38 of the attached document).

Did I mis-understand Dr. Shultz's original comment or am I reading the graphs wrong? I would greatly appreciate your help in resolving this matter.

Thanks very much,

Mark

Following this e-mail Drs. Jendrich and Schultz replied as follows:

Dear Mr. Kirk

After discussion of Mr. Schulz and myself we just want to give you some answers to your questions. In fact, there might be some misunderstanding about the through-wall attenuation.

We think there is no doubt that there is an attenuation of **fluence** through the wall and the fluence factor given in RG 1.99 Rev. 2 can be considered as conservative, since it gives a pretty flat curve, even flatter than most evaluations of dpa attenuation. Furthermore, we think that it reflects present knowledge that a corresponding decrease in **irradiation induced embrittlement within a homogeneous material** should be expected. The current IAEA programme (also cited in the EPRI report on page vi, NRC apparently planned some participation) to validate damage attenuation may result in a further step of verification of this. We also go along with Colin English et al. in that EPRI report that due to the large changes in spectrum you might rather use dpa than fast fluence as irradiation parameter. You just need a proper fluence - dpa correlation for the material in question.

However, regarding the post irradiation material properties of trepans of **real RPVs**, they show a different picture of the through-wall trends. They show that material inhomogeneity might strongly influence or even override the attenuation effect. This may be due to initial properties or uneven distribution of chemical elements enhancing embrittlement like Cu, P, Ni and Mn. And this might be the case particularly for the older vessels. This point of view may be a little more explicit, but basically in accordance with the conclusions of the EPRI report, see pages 3-17 and A-59.

Besides you mentioned the comparison of the Gundremmingen-A results with the RG 1.99 in KUS 90, figure 30. In fact this figure shows, that the RG 1.99, Rev.2 attenuation function is largely conservative. However, if the L-T orientation of the same trepan had been taken (see figure 15 of the same paper and pages 3-14ff of the EPRI report), the result would have been different, particularly if the starting point was the temperature shift at the depth of a ¼ wall thickness, i.e., the location, where the surveillance specimens usually come from. Again, we think to be fully in agreement with the conclusion of the EPRI report, that "in evaluating vessel embrittlement, the effect of any variations in start-of-life properties or chemical compositions must be taken into account". The question remains, of course, how to deal with this demand in case of a vessel in operation, where these data are usually not available.
Finally, there might be some (probably minor) effects, causing an unfavorable bias of the irradiation induced embrittlement near the outer surface of the RPV wall with respect to the inner part:

- Lower (irradiation) temperature near the outer surface.
- A possible flux effect, which might result in higher irradiation induced shift for the same fluence at lower flux. This effect is still under discussion, however, there were some substantial results presented recently at the last IAEA Specialists' Meeting (24 to 28 May 2004 at Gus Khrustalny (near Moscow)) supporting the hypothesis, that there is a significant effect for materials with Cu contents larger than some 0.15%. Unfortunately, the CD with the papers has not yet been distributed, so I cannot send you any papers. However, you might contact W. Server/ATI or R. Nanstad/ORNL, who also attended the meeting.

In conclusion, It might be physically justified to use an attenuation function for fluence or dpa to evaluate through-wall properties, since attenuation doubtless takes place. However, we still consider it reasonably careful to use constant through-wall properties (based on the surveillance programme results from the depth of a ¼ wall thickness) for the integrity analysis, because of the variability in through-wall properties shown by the results from trepans from real RPVs and the possible bias counteracting the fluence attenuation.

With best regards
Uwe Jendrich, jed@grs.de

Staff Response: Based on this email exchange it became clear that Dr. Schultz's comment did not pertain to nonconservatism in the Regulatory Guide 1.99 Revision 2 attenuation function itself, but to the potential nonconservatisms that could arise if use of that attenuation function (which Drs. Schultz and Jendrich state in their letter ***is*** conservative) attributable to the variability of chemical composition and initial unirradiated toughness through the vessel wall. In response, we would like to point out that the overall uncertainty in chemical composition and in initial unirradiated toughness is explicitly modeled in FAVOR, and that the magnitude of the uncertainty that we assume to be associated with these variables is larger than is characteristic of any individual weld, plate, or forging, so our treatment can be regarded as conservative (see Appendix D of [***EricksonKirk-PFM***] and Section 4.2.2.2 of [***EricksonKirk-SS***] for details). Additionally, FAVOR models the through-wall variability in chemical composition that can be characteristic of welds (see Section 4.3.2 of [***EricksonKirk-PFM***]). Finally, we note that the approach proposed by Drs. Schultz and Jendrich in their paragraph above that begins "in conclusion" is inconsistent with a best estimate approach (and unnecessarily conservative) because it intentionally ignores a measurable and well-recognized physical phenomena.

Reply to Reviewer Comment #74

Comment made by: Murley

Reply by: MEB

Comment: To better understand the differences between this study and the basis of the current PTS Rule it would be useful to have a comparative summary.

Staff Response: See Section 4.2.1 of this report. This section contains a summary of major differences between this investigation and the investigations that provided the basis of the current PTS Rule. This section indicates where these changes are discussed in detail, either in this report or in other supporting documents.

Reply to Reviewer Comment #75

Comment made by: Murley

Reply by: MEB

Comment: To better understand the PFM results it would be useful to see examples where the progress of a crack through the vessel wall is tracked.

Staff Response: See Appendix F of this report.

Comment made by: Murley

Reply by: MEB, PRAB, and SMSAB

Comment: What is the justification for selecting only 3000 seconds and 6000 seconds as the only possible reclosure times for transients involving stuck-open safety valves? Why is this rather coarse discretization an adequate representation of the continuum of possible events?

Staff Response: As detailed in Section 8.5.3.2 of this report, valve reclosure is a random event that can occur at any time after the transient begins. In our model we have discretized this continuum into two possibilities: reclosure at 3000 and 6000 seconds. These possibilities were selected recognizing that the severity of the transient varies with valve reclosure time. Up to some time transient severity will increase with increasing time before reclosure because the temperature of the primary system is dropping (which reduces the fracture toughness), while the thermal stresses are still climbing (because the cool-down is continuing). However, once the RCS has reached its minimum temperature (established by the temperature of the HPI water) the severity of the event will begin to reduce because the thermal stresses will begin to decline. The 6000-second reclosure time was selected to (approximately) coincide with the time of maximum transient severity because it is at (approximately) this time that the RCS temperature is reaching its minimum value. The potential for valve reclosure after very long times (in excess of 7200 seconds, or 2 hours) were not considered because by that time operators would have initiated new procedures. Since the operator's objective is to stop the transient (stop dumping irradiated primary system water into containment) they would likely depressurize the steam generators by opening the steam dump valves to cool the secondary side, and they would start low-pressure injection so that they could shut off HPI. These actions change the nature of the transient, making it more benign. Also they change the probability of operator error. The 3000-second reclosure time was selected because it is not reasonable to assume that all valve reclosures will occur at the worst possible time.

In response to this comment we performed a sensitivity study based on Palisades Transient 65 at 60 EFPY. Transient 65 (the most risk-significant stuck-open valve/reclosure case for Palisades) involves one stuck-open pressurizer safety relief valve that recloses at 6000 sec after the valve sticks open. Containment spray is assumed not to actuate, and no operator actions are modeled. In our sensitivity study we varied the valve reclosure time from 3000 seconds through 14000 seconds. The effect of these valve reclosure times on the conditional probability of crack initiation (CPI) and on the conditional probability of vessel failure (CPF) by through-wall cracking are illustrated in the figure on the following page. These results demonstrate the significant effect of varying reclosure time between 3000 and 6000 seconds where an increase in CPF of ~200-fold is seen. Conversely, the increase of CPF between 6000 seconds and the peak value is only an additional factor of ~2. Based on this information, use of our model where all possible valve reclosure times was discretized into two possibilities (reclosure at 3000 and 6000 seconds) was deemed adequate for the following reasons:

(1) The aim of this investigation is to provide a "best estimate" model, not a "worst case" model. As such, it is important that enough valve reclosure times be modeled to capture the effect shown in the graph below, not that the peak value be captured.

(2) The time of peak CPF is likely to vary slightly from transient to transient and from plant to plant due to differences in valve sizes, charging rates, etc. For some of the other cases modeled, CPF may peak closer to 6000 seconds than for the one particular case studied here in detail.

(3) As detailed in the first paragraph of the "staff response," valve reclosure at very long times need not be considered because these are effectively different transients than the stuck-open valve/reclosure case being modeled here.

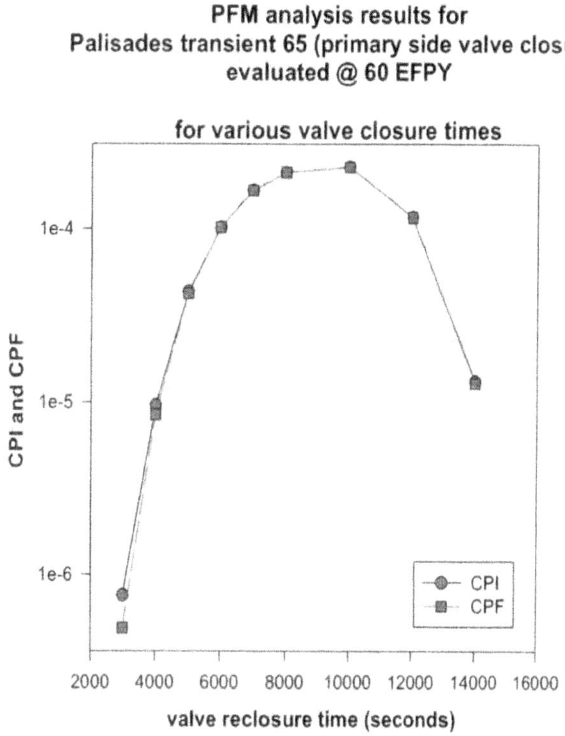

The following pages provide the final comments made by the review panel following receipt of the draft version of this report and all supporting documents (see Figure 4.1), and the staff's related responses. While the reviewers' final comments have been reformatted, they appear otherwise unaltered.

The portion of the reviewer's comments to which the staff is responding is indicated by highlighted text. The staff replies can be found in boxed text that follows the comment made by the reviewer.

Final comments made by reviewer Murley

PEER REVIEW OF THE PTS TECHNICAL BASIS
Thomas E. Murley
Prepared for NRC Office of Research
November, 2004

1. INTRODUCTION

This report summarizes my comments based on a review of most of the large amount of PTS material provided to the peer review panel.

The NRC research staff is to be congratulated for producing the breadth and quality of world class PTS research represented by this material. Specifically, there are greatly expanded PRA analyses of PTS events, comprehensive thermal-hydraulic calculations of the many classes of PTS transients, improved PFM methods and much new materials data, especially on flaw distributions and flaw sizes in reactor pressure vessels.

The issue of pressurized thermal shock is one of the most complex safety issues to analyze for light water reactors, because it involves the integration of several technical disciplines and safety questions, and the ultimate failure mode of a through-wall vessel crack has never been experienced in an operating reactor pressure vessel. It is clear that today's understanding of PTS phenomena is much greater than when the original PTS analyses were done in the early 1980's. Current estimates of the RPV failure probability suggest that they are substantially lower than thought twenty years ago. The NRC research staff's primary conclusion is that the PTS regulation (10 CFR 50.61) has a large degree of conservatism and that current methods and data support the potential relaxation of the regulation to remove some of the unnecessary conservatism. I agree with that conclusion.

The staff has chosen for its safety metric the surrogate parameter through-wall crack frequency (TWCF) which they equate to core damage frequency. The logic of their overall approach to calculating TWCF is shown in Figure 3.1 of draft NUREG-1806, and I find that approach to be reasonable. The main blocks of that approach are (a) PRA Event Sequence Analysis, (b) Thermal-Hydraulic Analysis, and (c) Probabilistic Fracture Analysis. While I have some issues and concerns with each of these areas of analysis, these concerns do not rise to the level that would seriously challenge the logic of the overall approach or the general validity of the PRA, TH or PFM calculational methods. Rather, I believe these concerns can best be dealt with by a conservative approach to revising the PTS Rule. This is discussed later in my report.

> <u>Staff response</u>: In Chapter 10, we conservatively equate through-wall cracking with a large early release, not with core damage.

As one might expect in a comprehensive analysis of an issue as complex as pressurized thermal shock, there remain areas of uncertainty, known conservatisms and apparent nonconservatisms. But reasonable regulatory assurance does not require absolute assurance, and NRC has regulated nuclear reactor safety over the years in the face of many kinds of technical uncertainties. Accordingly, I believe this PTS analysis can serve as the basis for further regulatory action to revise 10 CFR 50.61.

I understand the purpose of this peer review is to help the NRC research staff produce a robust technical basis for any changes in the current PTS Rule. Therefore, the peer review panel can have particular value in looking for flawed analyses and overlooked phenomena. It is in that spirit that many of the comments in the following sections are directed at areas where there may be errors, gaps or nonconservatisms in the analysis.

2. PRA EVENT SEQUENCE ANALYSIS

The safety risk posed by pressurized thermal shock of reactor pressure vessels was not known at the time WASH-1400 was published in 1975. Later, when the PTS Rule was developed in the early 1980's, there were some rudimentary probabilistic calculations made but the staff at that time cautioned against using them to derive licensing requirements. The large uncertainties in probabilistic evaluations at the time led the staff to use them to estimate the level of safety rather than attempt to derive licensing requirements from the probabilistic results. Subsequent to the rule being published, there was a general feeling that as long as a plant's reactor vessel was below the screening criteria the TWCF was acceptably low and, therefore, detailed PRA studies of PTS risk were not warranted.

The current NRC reevaluation of the PTS Rule is therefore the most comprehensive attempt to identify and analyze potential PTS event sequences. It is an impressive achievement. Thousands of event sequences were analyzed, including those at hot shutdown conditions, and they included a detailed, realistic treatment of operator actions.

At the May meeting of the Peer Review Group we were told that the scenario boundary conditions were more important than uncertainties in the TH models in RELAP5 for calculating TWCF. Some of these important boundary conditions are break size, break location, power level (decay heat load), seasonal effects (temperature of HPI water), number of valves sticking open, valve reclosure time, and timing of operator actions (throttling of HPI). These boundary conditions are products of the PRA analysis and are input to RELAP5. They generally seemed reasonable to me with only a couple of questions below.

The staff used plausibility arguments to reduce the number of transients to a manageable number and the description of the binning process seems to me to be logical and either conservative or neutral with regard to estimating TWCF.

In order to judge whether the PRA results for Oconee 1, Beaver Valley 1 and Palisades were representative of other PWRs, the staff conducted a qualitative examination of the important design and operational characteristics of five other PWRs to judge their susceptibility to PTS transients. The approach is logical and their conclusion that "the TWCF estimates produced for the detailed analysis plants are sufficient to characterize the TWCF estimates for the remainder of the PWRs" seems reasonable to me.

A much less rigorous (but thought to be conservative) bounding risk analysis of external event contributors to TWCF was also carried out. The conclusion is that external event contributors (principally seismic and fires) are not greater than for the internal PTS event sequences analyzed and that a realistic analysis would likely show external event contributors to be substantially less than for internal events. This conclusion seems reasonable, but the lack of a definitive analysis for external events is a residual uncertainty that will be discussed later in this report.

Regarding details of the PRA analysis, on page 5-64 it is assumed that the SRV opening size is uniformly distributed (any specific opening is equally likely). This assumption seems intuitively wrong. Why is there not a peak in the distribution at the fully open position? It seems logical to me that the force opening the SRV would most often drive it to the fully open position where it would stick open and only rarely would the valve stick open at some intermediate position. If this is correct, the PRA assumption would be nonconservative. What do the valve experts say?

Staff response: The basis for the assumption questioned by reviewer Murley was reexamined, and the staff determined that the most likely scenario would be for the valve to fully open when demanded (as suggested by reviewer Murley). The value of the basic event representing the valve opening was set to 1.0 (representing a full open condition), and the Beaver Valley and Oconee PRA models were re-quantified (this issue does not exist in the Palisades model). The new PRA distribution information was then post-processed with the FAVOR conditional through-wall crack probability results, yielding the following revised total TWCFs:

Plant	EFPY	Original TWCF	Revised TWCF	Revised/Original TWCF
Oconee	32	2.30E-11	7.43e-11	3.23
	60	6.47E-11	2.06e-10	3.18
	Ext-Oa	1.30E-09	2.75e-9	2.12
	Ext-Ob	1.16E-08	1.81e-8	1.56
Beaver Valley	32	8.89e-10	1.36e-9	1.53
	60	4.84e-9	6.12e-9	1.26
	Ext-Ba	2.02e-8	2.14e-8	1.06
	Ext-Bb	3.00e-7	3.05e-7	1.02

The following observations may be made regarding these results:

- These results are consistent between the plants. The apparent effect is higher at Oconee at lower EFPY because at 32 and at 60 EFPY Oconee is at a much lower embrittlement than Beaver. Consequently, the Oconee results at 32 and at 60 are more dominated by changes to the stuck-open valve modeling, so increasing the severity of these transients have a bigger result on the overall outcome. Oconee and Beaver have (roughly) equivalent embrittlement at (respectively) Ext-B and 60 EFPY. At these EFPY the results in the tables above show (respectively) a factor of 1.56 or 1.26 increase.
- The revised (higher) TWCFs are still small numbers (E-10 to E-9 at 60 EFPY), and as such are not expected to alter the overall conclusion that sufficient margin exists to support a revised PTS Rule.

Current plans do not call for a revision of NUREG-1806 to incorporate this change into the baseline model (due to the small effect on the overall results). However, this new information may be used in future revisions to NUREG 1806.

In Appendix B the staff justifies the model where all possible valve reclosure times were discretized into two possibilities (reclosure at 3,000 and 6,000 seconds). Those sequences where a valve or valves reclose at 3,000 seconds contribute very little to TWCF, while valve closure sequences at 6,000 seconds contribute to TWCF over 100 times greater. Later closure times make a contribution to TWCF that is more than a factor of 2 greater than closure at 6,000 seconds. It is not clear whether a finer discretization of reclosure times would yield less conservative results (i.e., a greater estimated TWCF). I regard this as another residual uncertainty.

Staff response: When considering the information presented in "Reply to Reviewer Comment 76," it is important to recognize that the aim of our discretization of the infinity of possible value reclosure times into two discrete possibilities (reclosure at 3000 seconds and reclosure at 6000 seconds) is not to select the reclosure time that produces the highest possible CPTWC (as seems to concern reviewer Murley) but rather to capture the variation of CPTWC with valve reclosure time. Thus, we are effectively trying to represent the area under the curve shown in Comment 76 as a discrete number of steps. As illustrated by the figure below, representing all possible reclosure times just with reclosures at 3000 and 6000 seconds achieves this goal. Additionally, it should be noted that reviewer Johnson's views differ from reviewer Murley's views on this point; reviewer Johnson views this aspect of our model as being, if anything, unnecessarily conservative.

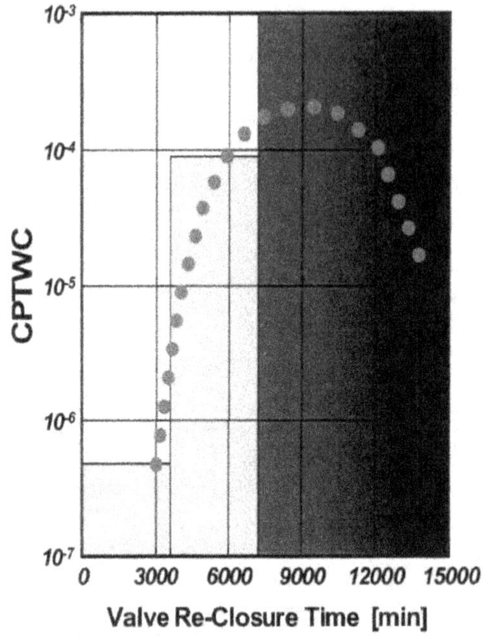

3. THERMAL HYDRAULIC ANALYSIS

The thermal-hydraulic module of the PTS analysis takes the skeletal outline of event sequences and boundary conditions from the PRA module and provides time-dependent downcomer fluid temperature, pressure and heat transfer coefficients to the probabilistic fracture mechanics (PFM) module. Over five hundred RELAP5 calculations of pressurized thermal shock transients were carried out.

To validate the results from RELAP5 an extensive assessment was made against experimental data from separate effects and integral system tests. RELAP5 was generally able to predict system pressure and flows well, but prediction of fluid temperature in the downcomer was more problematic. Sensitivity studies of downcomer heat transfer shown in Table 9.6 shows that RELAP5 with Petukhov-Gnielinski fluid to wall heat transfer modeling predicted TWCF of 4.5 times greater than the baseline RELAP5 calculations when averaged over 12 transients. I regard this as another residual uncertainty.

As a result of these studies, the staff concluded that RELAP5 "is capable of well-predicting the phenomena of importance for evaluating PTS risk in PWRs."

The bases for the staff's conclusions appear to be that (a) the initial and boundary condition of the transients from the PRA module are far more important than the details of RELAP5 models, (b) the one-dimensional RELAP5 calculations of system pressure and flows cannot be too far from reality, and (c) in any case it doesn't make much difference to the final estimate of TWCF because there is good fluid mixing in the downcomer and the temperature of cold water in the downcomer dominates the determination of the vessel wall temperature. These arguments are superficially plausible, and they may even be correct, but I still find them somehow unsatisfying.

In the late 1970's NRC invested several million dollars in developing and validating a multidimensional systems code (TRAC) that was intended to be the benchmark tool for thermal-hydraulic analyses of complex LWR transients and accidents. It was recognized at the time that TRAC would be expensive to maintain, but it was believed to be worth the expense to have available such a benchmark tool. I have lost touch with the status of TRAC over the years, but nonetheless I find it disappointing that no benchmark multidimensional calculations were done in order to help answer some of the TH questions concerning this PTS evaluation, particularly about fluid conditions in the downcomer and heat transfer between the fluid and vessel wall.

4. PROBABILISTIC FRACTURE MECHANICS ANALYSIS

The area of PTS analysis that has made the greatest strides in the past decade is probabilistic fracture mechanics. The amount of new data and the depth of analysis are truly impressive. It seems to me that the new analyses cited as reasons for suggesting the PTS Rule is conservative are solidly based. By far the biggest factor contributing to the reduced estimates of reactor vessel failure is the new data on vessel flaw distributions. The estimated TWCF drops by a factor of between 20 and 70 when the new flaw distribution is adopted instead of the older Marshall distribution used in the early 1980's.

The FAVOR code has undergone an extensive V and V program to demonstrate that the software actually calculates what the theory intends. It has been concluded that FAVOR "meets the requirements stated in the theory manual and the user's guide with reasonable confidence in the accuracy of the FAVOR-generated results." From my review it appears that this software validation program has been thorough and that the conclusion above is valid. One may, however, have questions about the theory itself.

One must acknowledge that the FAVOR code represents an excellent technical achievement. There are logical treatments of the physical and mechanical processes through complex models, and the mathematical treatments are elegant. Still, we must also be aware that engineering judgment was used for dozens of choices made throughout the development of the models. One can imagine an alternate world, with equally brilliant engineers faced with the identical data and uncertainties and using comparably **(1)** elegant mathematics, where a different computer code with different models would emerge. Who could say which one would be a better representation of reality? The point of this flight of fancy is not to suggest a new PFM code but to point out that FAVOR's results have an unknown degree of uncertainty simply from the engineering judgments that went into its models. During my review I have seen the results from three different versions of FAVOR, and there have been revisions to the models in recent **(2)** months. A bug was identified in how FAVOR associated material properties with cracks that lie on the fusion line of welds. In response to Peer Review comments FAVOR was modified to implement a new upper shelf model and to include the effect of pressure acting on the crack surfaces.

Staff response:

Regarding Comment 1: Given that the materials and fracture experts on both the external review panel and on the ACRS have never raised any serious concerns regarding either the overall PFM model or its specifics, it seems unlikely that reviewer Murley's "alternate world" exists. Additionally, it may be noted that the various aspects of the PFM model have been the topics of multiple presentations at public meetings, as well as of presentations made to professional societies (both domestically and internationally), and of publications in peer reviewed journals. Serious concerns regarding the theoretical underpinnings or the implementation of the PFM model have not arisen in any of these forums.

Regarding Comment 2: Reviewer Murley is correct in pointing out that the specifics of the FAVOR code, as well as the specifics of the overall PTS model, have experienced multiple evolutions from that reported in [***Kirk 12-02***]. These changes have been motivated by the comments the staff has received from the many individuals and organizations who have reviewed our work (i.e., the ACRS, this review committee, the commercial nuclear power industry (EPRI/MRP), and the Office of Nuclear Regulatory Research) as well as by efforts on the part of both the staff and its contractors to check and improve our work. We believe that these efforts to review and improve our model should provide increased confidence in the robustness of our results rather than providing cause to question them. Additionally, it should be pointed out that the net effect of all the changes made to the model and input data on the results has been small. The TWCF values reported herein are the following percentages of those reported in [***Kirk 12-02***],

For Oconee, the mean TWCF in this report is ≈5% of that reported in [***Kirk 12-02***].
For Beaver Valley, the mean TWCF in this report is ≈150% of that reported in [***Kirk 12-02***].
For Palisades, the mean TWCF in this report is ≈50% of that reported in [***Kirk 12-02***].

The large reduction in the Oconee TWCF resulted from removal of an error in the thermal-hydraulics model of the 16-in. diameter break, while the increase of 50% in the Beaver Valley TWCF values resulted from correction of an error in the FAVOR code (commented on by reviewer Murley) that was revealed as part of our V&V process. Neither of these changes affected the Palisades results; the 50% reduction in TWCF reported above for Palisades therefore best represents the combined effect of all of the model changes made between the issuance of [***Kirk 12-02***] and this report.

A major question unanswered is how well the flaw data represent all US PWR pressure vessels. The staff used the Shoreham vessel flaw data, thought to be conservative, but concluded that "it is not possible to ensure on an empirical basis alone that the flaw distributions based on these data apply to all PWRs in general."

Staff response: Following the quotation provided by reviewer Murley the staff goes on to state (in Section 7.5 of this report) that *"flaw distributions proposed in [**Simonen**] rely on the experimental evidence gained from inspections of the materials summarized in Table 7.1 and do not rest solely on this empirical evidence. Along with these data Simonen et al. used both physical models and expert opinions when developing their recommended flaw distributions. Additionally, where detailed information was lacking Simonen et al. made conservative judgments (for example, all NDE indications were modeled as cracks and, therefore, potentially deleterious to RPV integrity). This combined use of empirical evidence, physical models, expert opinions, and conservative judgments allowed Simonen et al. to propose flaw distributions for use in FAVOR that are believed to be appropriate/conservative representations of the flaw population existing in PWRs in general."* For the stated reasons the staff's views on the general applicability of the flaw distribution are not nearly as bleak as one would surmise based on the limited quotation (which was not stated as a conclusion in any of the documentation supplied to the review group) provided by reviewer Murley.

The staff proposes that the reference temperature for axial welds, circumferential welds and plates be calculated from Equations 8-1, 8-2 and 8-3, respectively. As a thought experiment let us suppose the maximum values of the reference temperature of axial welds, circumferential welds and plates are identical, say RT (peak). The equations above would yield

$$RT_{CW} = RT\ (peak)$$
$$RT_{PL} = RT\ (peak)$$
$$RT_{AW} < RT\ (peak)$$

The equation 8-1 averages the reference temperature of the axial welds in the high fluence mid region of the vessel with the reference temperature of upper and lower axial welds in the low fluence regions of the vessel. The resultant RT_{AW} is an average value that is lower, and may be substantially lower, than the peak value of reference temperature in an axial weld. Thus, there seems to be an inconsistent treatment of axial welds relative to plates and circumferential welds when calculating reference temperatures. One could argue that the formulae for RT_{PL} and RT_{CW} will always yield conservative results. However, the formula for RT_{AW} may be nonconservative, because the averaging process could mask a single highly embrittled axial weld.

Staff response: The situation that reviewer Murley proposes in his thought experiment cannot happen. According to Eq. 8-1, which prescribes the method for calculating RT_{AW}, the reference temperature associated with each axial weld fusion line from which the weighted metric RT_{AW} is calculated is the *maximum fluence* occurring along the axial weld fusion line. As such, there is no influence of the lower fluence regions at the edge of the vessel beltline on RT_{AW}. Additionally, as stated in Section 8.4.1, it is not important that the reference temperature metrics used to correlate the TWCF results reflect the peak reference temperature in each region of the vessel, but rather that the reference temperature metrics reflect the reference temperature of the vessel steel at the location of postulated flaws.

Reviewer Murley's comment caused the staff to re-examine its proposed reference temperature metric formulae. This reexamination revealed that the proposal in the draft of NUREG-1806 that reviewer Murley received produced an inconsistent treatment of the TWCF contribution of axial welds *vs.* that of both plates and circumferential welds. The proposal in the draft version of NUREG-1806 featured an axial weld metric based on an average of maximum RTs occurring along each fusion line weighted to account for differences in weld fusion line length (and thereby number of simulated flaws). Conversely, the draft version of NUREG-1806 featured plate and circumferential weld metrics that were not weighted to account for differences in plate volume or circumferential weld fusion line length (and thereby number of simulated flaws). Our discussion of reference temperature metrics and their use to develop reference temperature screening limits in Section 11.4 has therefore been modified to produce a more consistent treatment across different material regions (i.e., axial welds, circumferential welds, plates, and forgings). In this section, both weighted average RT metrics and maximum RT metrics are developed, and both are used to proposed new RT screening limits for PTS.

In welds a gradient of properties through the thickness of the vessel is expected to exist because of changes in chemistry content of the various weld layers. FAVOR adopts a 4-weld layer model wherein the chemistry is re-sampled from a distribution every time a crack passes t/4, t/2, and 3t/4 in the vessel wall. Data on the plants indicate that Palisades has three axial weld layers and eight circumferential weld layers, Beaver Valley 1 has two axial weld layers and seven circumferential weld layers, and Oconee 1 has two and three axial weld layers and seven circumferential weld layers. Clearly a 4-weld layer model does not represent any of these welds. Developing a more complex weld layer model would have required significant changes to FAVOR and therefore was not done.

The staff report states that the 4-weld layer model reduces the estimated TWCF by a factor of 2.5 relative to a model that has a constant mean value of chemistry through the weld. It is not immediately obvious why a 4-weld layer model should reduce the estimated TWCF, unless the sampling logic somehow discriminates against crack arrest.

Staff response: When chemistry is resampled, there is an opportunity to simulate either a more or a less radiation sensitive material than existed in the preceding weld layer. If a more radiation-sensitive material is simulated, the crack will most likely continue to propagate. However, if a less radiation-sensitive material is simulated, the toughness will increase as the crack moves into the next weld layer, making arrest more likely. This increases opportunity for arrest would not exist in a PFM model without weld layers. As discussed in NUREG-1807, the existence of distinctly different copper levels through the thickness of vessel welds is well documented, providing a physically plausible basis for the increased arrest capability simulated by the FAVOR code.

The sampling protocol in FAVOR requires estimated chemistry (Cu, Ni and P) values for each weld and plate subregion of the vessel. The sampling protocols distinguish between the first flaw simulated in a subregion and all subsequent flaws in the subregion. The standard deviations for chemistry content when sampling for subsequent flaws are much less than the standard deviations when sampling for the first flaw. This logic seems odd to me. Apparently the thought is that Flaw 1 has established a new chemistry reality for that subregion of the vessel, that this new chemistry reality is governed by local variability parameters, and that all subsequent flaws must be governed by the new chemistry reality established by Flaw 1. But the fact is that the actual chemistry of a subregion is unknown (represented by a distribution) and that subsequent flaws should be subject to the same chemistry distribution as the first flaw.

Staff response: If two flaws are simulated to exist close to each other (i.e., within the same sub-region) in the same vessel, it is not physically possible that the differences between the chemistry values associated with these two flaws will be as variable as the differences between the chemistry values associated with two flaws that have a large spatial separation.

The point of these comments is to illustrate that there are residual uncertainties in the PFM calculations, just as there are in the PRA and TH calculations.

5. ACCEPTANCE CRITERION

In Chapter 10 the staff addresses possible risk-informed reactor vessel failure frequency acceptance criteria. In order to be consistent with NRC's Safety Goal policy and other policies the staff had to consider sequences beyond vessel cracking to include severe fuel damage, fission product release and containment failure. The Accident Progression Event Tree (APET) in Figure 10.1 seems logical and thorough, but the analytical and experimental bases to quantify the many branches of the event tree do not exist.

The staff has chosen the definition of reactor pressure vessel failure to be when a PTS-induced crack penetrates the vessel wall, and not when a PTS event initiates a crack in the vessel wall. I agree with the staff's choice. The phenomenon of crack arrest has been well demonstrated in tests over the years. The staff has also assumed that a through-wall crack equals core damage which is assumed to equal a release of radioactivity. This is acknowledged to be conservative, but for purposes of defining a reactor vessel failure frequency acceptance criterion it is a reasonable assumption.

Much less study of the consequences of vessel failure accidents has been done than is the case for core damage accidents resulting from undercooling or ATWS events. As a result, the question arises whether vessel failure accidents could lead to especially large early release scenarios. In particular, the ACRS has raised the issue of potential Large Early Release Frequency (LERF) source terms from air oxidation of fuel in some of the most severe (and unlikely) vessel failure scenarios. I do not think it would be a wise use of resources to mount a substantial research effort to try to answer all the questions surrounding air oxidation source terms.

The staff makes a reasonable case that the conditional probability of a large early release of radioactivity, given a PTS-induced vessel failure, is small (less than 0.1) to extremely small (much less than 0.01). Based on their largely qualitative analysis the staff suggests an acceptance criterion of TWCF = 1E-6/r-y.

One can reach a similar conclusion on acceptance criterion via a different line of reasoning. The current acceptance criterion in Reg. Guide 1.154 is TWCF = 5E-6/r-y. However, research has shown that the methods used to estimate TWCF at the time Reg. Guide 1.154 was written are highly conservative. Thus, it would be easier for a vessel today to demonstrate compliance with TWCF = 5E-6/r-y than it would have been for the identical vessel to have demonstrated compliance fifteen years ago, say. Viewed in this way it would be reasonable to lower the acceptance criterion below
5E-6/r-y if one wanted to maintain comparable margins using current methods and data for calculating TWCF.

The staff's reasoning is probably sounder, and in any case I agree with an acceptance criterion of TWCF = 1E-6/r-y.

6. RESIDUAL UNCERTAINTIES

This PTS analysis is described by the staff as a "best estimate analysis" but the fact is that *there are dozens of instances where engineering judgment is used in place of detailed analysis*. In most of these instances the engineering judgments are claimed to be conservative, and they may well be, but there is no corresponding discussion of potential nonconservatisms and residual uncertainties. Throughout the report there are repeated assertions of conservatism in the calculation, whereas several instances of nonconservatisms of factors of ~2 are dismissed as not significant. While this may be true for individual cases, we must keep in mind that these nonconservatisms and residual uncertainties are cumulative.
To illustrate the point, I have listed below some of the issues I found in my review and their potential impact in increasing TWCF.

External events (up to a factor of 2)
SRV opening size (unknown)
Valve reclosure times (up to a factor of 2)
Downcomer heat transfer (up to a factor of 4.5)
Applicability of flaw data to all PWRs (unknown, but could be large)
Method of calculating RT_{AW} (unknown)
4-weld layer model (up to factor of 2.5)

I recommend that the staff prepare a similar but more comprehensive list of all potential conservatisms, nonconservatisms and residual uncertainties in the analyses, before embarking on rulemaking, in order to try to get a clearer picture of the overall uncertainties in the calculations of TWCF.

> Staff response: In general, reviewer Murley's comment that a more balanced perspective of the conservatisms *vs.* nonconservatisms that exist in the model is well taken. For this reason, we have revised and expanded upon the discussion found in Section 11.4.3 of this report.
>
> With regard to the specific uncertainties reviewer Murley details above, specific staff replies can be found as boxed text earlier in this letter.

7. APPROACH TO RULEMAKING

From my review it is clear that NRC is faced with the situation where a great deal of high quality PTS research has demonstrated that the current PTS Rule has a large degree of conservatism, but the analysis itself has a host of residual uncertainties. The traditional way NRC has dealt with such situations in the past has been through conservative decision making. I believe that is the best approach in this case.

It appears that the research staff is proposing an approach to rulemaking that would leave the basic form of the current PTS Rule (10 CFR 50.61) intact. In Chapter 11 the staff proposes reference temperature based PTS screening criteria along with methods for calculating the reference temperatures. I agree with this general approach, but I believe the locus method for calculating acceptable reference temperature screening limits is unnecessarily complicated. From Figures 11-4 and 11-6 one can deduce the following proposed (approximate) reference temperature screening limits:

For axial welds	$RT_{AW} \leq 300°F$
For plates and forgings	$RT_{PL} \leq 375°F$
For circumferential welds	$RT_{CW} \leq 455°F$

Because of the residual uncertainties discussed in the last section, I believe all of these screening limits are too high. The case of circumferential welds requires special attention because the frequency of crack initiation for circumferential welds may be several times greater than for axial welds or plates (over 30 times greater for Beaver Valley 1). There may be operational reasons why $RT_{CW} = 455°F$ (a very highly embrittled weld) should not be allowed. If there were an overcooling event where a flaw in a highly embrittled circumferential weld initiated a crack that subsequently arrested, the NRC for safety reasons would certainly not permit such a vessel to go back into service without repair. In practice such an event would mean that the plant's useful service life would be at an end.

The staff asserts that, because some of the uncertainties in the 1982 analysis have been reduced and have been considered explicitly in the current PTS models, the use of margins in the proposed reference temperature screening limits is inappropriate. I disagree with that assertion.

> Staff response: In principle we agree with reviewer Murley's view that the appropriateness (or not) of adopting the proposed *RT*-based screening limits without additional margin can be assessed by considering the balance between the conservatisms *vs.* the nonconservatisms that remain in the model used to develop the screening limits. Section 11.4.3 of this report presents such a comparison which shows that more conservatisms than nonconservatisms remain in the model. Consequently, the staff stands by its recommendation that the proposed *RT*-based screening limits can be used without additional margin.

In view of the residual uncertainties discussed earlier, and there are no doubt other uncertainties, the **(1)**
traditional regulatory approach to conservative decision making using margins is appropriate for this situation. After preparing a comprehensive list of all potential conservatisms, nonconservatisms and residual uncertainties in the current analyses the staff will be in a better position to judge how much margin, if any, is appropriate. My own preliminary judgment on an appropriate set of reference **(2)**
temperature screening limits would be the following:

For axial welds	$RT_{AW} \leq 290°F$
For plates and forgings	$RT_{PL} \leq 350°F$
For circumferential welds	$RT_{CW} \leq 350°F$

> Staff response:

Regarding comment (1): Here, reviewer Murley raises the question of incompleteness uncertainty, a topic discussed in Section 3.2.2.5 of this report where we describe the process employed to ensure that incompleteness uncertainty is explicitly examined and (hopefully) held to a minimum. Since 2002, the results of this project have been reviewed by four different independent groups, as well as by the authors themselves. Revisions to, improvements in, and additions to the model have been made over the past 2 years to address the comments raised and errors found by these various groups. As mentioned earlier, the net effect of all of these changes has been a 2x reduction in the reported values of mean TWCF. Certainly, incompleteness uncertainty is a factor to be considered by decision makers using the results of this study, however the processes used to address incompleteness uncertainty and the small net effect of model changes on the TWCF results over 2 years time should also be considered.

Regarding comment (2): In the first sentence under item 6 (highlighted in *red italic*), reviewer Murley criticizes the approach of using engineering judgment in place of detailed analysis. Yet here, reviewer Murley appears to employ his own engineering judgment to establish margins on the proposed *RT*-based screening limits. Reviewer Murley's proposed margins are markedly different for axial welds (10°F) than for plates (25°F) than for circumferential welds (155°F), yet a technical justification beyond "judgment" is not forwarded by the reviewer. In his proposal regarding margins, reviewer Murley seems to violate his own guidance.

8. REGULATORY RESPONSE TO OVERCOOLING EVENTS

Table 8.5 shows that the frequency of crack initiation is much greater than the frequency of through-wall cracks. For Beaver Valley 1 the actual value of circumferential weld crack initiation frequency is about 5E-7/r-y. These figures demonstrate that the probability of a PTS-induced crack initiation and subsequent crack arrest, while low, is not negligible. It was this observation that led me to a comment in my preliminary report:

" . . . a question arose that may represent a gap in the current regulatory fabric governing pressurized thermal shock. What are the regulatory requirements for a plant that has suffered a severe overcooling event where the vessel did not have a thru-wall crack and no outward sign of damage but may have suffered a crack initiation that subsequently arrested in the vessel wall? How would the PTS risks change if such a cracked vessel went back into service? What criteria would NRC use to judge whether a complete inspection of the vessel was needed after a severe overcooling event? What inspection techniques would be required and what would be the scope of such inspections?"

The staff responded to these questions in Appendix B, page B-48. Their conclusion was that a severe overcooling event would violate the facility pressure-temperature limits in the licensee's technical specifications and that would require reporting to NRC and an evaluation if the reactor coolant system is acceptable for continued operation. Such evaluations by the licensee would likely follow ASME Code Section XI, Appendices A and E, which could lead to the inspection of relevant portions of the vessel. I find the staff's response to be an acceptable answer to my earlier questions.

9. CONCLUSIONS

Based upon my review of the PTS Technical Basis reports, I have reached the following major conclusions:

The current PTS regulation has a large degree of conservatism, and current methods and data support the potential relaxation of the regulation to remove some of the unnecessary conservatism.

The NRC research staff's overall approach to estimating Through Wall Crack Frequency is logical and well carried out.

The FAVOR V and V program was thorough and the conclusion that the code meets the requirements in the theory manual is valid.

The staff's proposed acceptance criterion for reactor vessel failure frequency of TWCF = 1E-6/r-y is reasonable and appropriate.

Even after this prodigious analysis effort there remain many areas of uncertainty, known conservatisms and apparent nonconservatisms.

I disagree with the staff's assertion that the use of margins in the proposed reference temperature screening limits is inappropriate.

The NRC can deal with the residual PTS uncertainties through conservative decision making, just as they have dealt with similar technical uncertainties in the past. This is not inconsistent with the principles of risk-informed regulation, which I support.

> Staff response: See staff replies to these concerns in the main body of reviewer Murley's comments.

10. RECOMMENDATIONS

I believe there will need to be continued NRC research on PTS and pressure vessel integrity for the foreseeable future. These are the main areas where I see the need for additional research:

In the near term, before embarking on rulemaking, the staff should make a comprehensive list of all potential conservatisms, nonconservatisms and residual uncertainties identified by the staff and peer reviewers, in order to try to get a clearer picture of the overall uncertainties in the PTS analyses.

The staff should begin planning to revise Regulatory Guide 1.154, since it will no doubt be an integral feature of any revised PTS regulation.

The staff should arrange to maintain a multidimensional systems code (or set of codes) for benchmark calculations of LWR safety issues.

This peer review effort has made clear to me that FAVOR still faces maturation as a reliable PFM code. I believe that only a few people in the world are familiar with the details of the models in the code and are capable of using it comfortably. I recommend that NRC support an international FAVOR Users Group to examine its logical structure, its models and assumptions and to test its output against as wide a range of test data as possible.

11. Editorial Comments:

- When this report is finalized and published it likely will be the definitive PTS technical basis document world-wide for years to come. Therefore, it is important for the NRC research staff to take the time to assure it is of the highest quality. In my review I encountered many editorial and grammatical errors as well as areas that are confusing or unclear. The following are some of the editorial comments:

- Page iii and page 1-12: The screening limit of 5E-6 was not discussed in SECY 82-465. The discussion on page 2-19 gets it right.
- Page 6-67: PIRT is Phenomena Identification and Ranking Table.
- Section 6.8.2.2: Second paragraph is garbled.
- Figure 7.4 is missing.
- Page 7-122: Is it really correct that crack arrest toughness and upper shelf toughness are uninfluenced by irradiation?

Staff response: The statement made is that the *temperature dependency* of crack arrest and of upper shelf toughness is not influenced by irradiation. The claim was never made that the actual toughness values are influenced by irradiation because, in fact, they are.

- Figure 7.6: Where is RT_{NDT}* defined?
- Page 8-143: Table 8.4 does not list 48 EFPY.

Staff response: This is correct; analyses were not conducted at 48 EFPY. TWCF values at 48 EFPY discussed in the text were arrived at by interpolating the 32 and 60 EFPY results.

- Table 8.4: RT_{PTS} is a 10 CFR 50.61 term – how is it defined here?
- Figures 8.3 and 8.4: Do not at all illustrate (to me) what they are purported to illustrate.
- Page 8-146: nfl – remove "circumferential welds"
- I_{FL} – remove number of plates
- Figure 8.6: K_I is not defined – appears to be applied K_I
- Figures 8.12-8.17, 8.28-8.31, 8.34-8.38, 8.42, and 8.44-8.47 are not readable.
- Page 8-168, item 1: Thickness should be diameter.
- Section 8.5.4.1: Is it 250°F or 260°F?
- Table 9.4: 130% htc – are values plus or minus?
- Page 9-239 and 12-269: Accepting should be Excepting.
- Section 11.4.1: Item 3 is garbled.
- Page F-174, Figure 7.3: Why wasn't K_{Ia} resampled at t/4, t/2 and 3t/4?

Staff response: As described in [***EricksonKirk-PFM***], K_{Ia} is never re-sampled at the t/4, t/2 and 3t/4 locations: chemistry is resampled, from which a new reference temperature is calculated, from which a new distribution of K_{Ia} is calculated. Having said this, chemistry is only resampled when the flaw is propagating in a weld because only welds are subjected to systematic through-thickness chemistry gradients. In Figure 7.3, chemistry is not resampled because in that figure, the crack is propagating through a plate.

Final comments made by reviewer Schultz

Peer Review
Technical Basis for Revision of the Pressurized
Thermal Shock (PTS) Screening Limit in the PTS
Rule (10 CFR 50.61): NUREG 1806
Comments by Helmut Schulz
Gesellschaft für Anlagen- und Reaktorsicherheit (GRS) mbH
Head of Component Integrity Department
November 25, 2004-11-25

1. Foreword

As a non US member of the review panel being not so familiar with the technical positions within the US nuclear community my views may reflect the differences in the background compared to other panel members. Since my expertise is in the field of nuclear safety assessment in general, fracture mechanics and component integrity specifically my comments are restricted mainly to these subjects.
The time available for the review of the final reports did not allow an in depth review of all the extensive material supplied by the NRC staff, therefore my comments are mainly based on the summary report.

2. Aspects of the overall approach

2.1 Structure and scope of the study

The study presented is well structured and the total effort is impressive. The report is written in a consistent manner identifying references as necessary.

The scope of the study covers all event sequences in the range from zero power hot stand-by up to 100% power. As stated in the appendices (page 13-53) low temperature over-pressure conditions are treated as a separate subject.

2.2 Regulatory frame

The regulatory frame is the risk-informed regulation which is practiced in the US. Most other countries have not yet applied risk-informed regulation in a similar advanced state therefore it is not meaningful to compare the methodology applied and results achieved in the study to existing regulations in other countries.

Because a number of countries follow the US regulation it has to be expected that this study will have a broad influence on technical positions regarding PTS.

Considering the defense-in-depth principle the annealing of the reactor pressure vessel to reduce considerably the radiation induced embrittlement of the material may provide a more robust solution compared to a risk-based approach which relies on a broad range of well developed capabilities.

2.3 Methodology

The probabilistic risk assessment, human reliability analysis and thermal-hydraulic analysis are not my expertise, although I would like to mention two items which may be already treated by other reviewers.

- Item 1: Section 8.5 of the summary report shows that the stuck-open valve transients with reclosure are a major contributor to the overall results. It is the view of the reviewer that valves are a sensitive area for maintenance errors which are difficult to treat in a probabilistic analysis. The numbers which are based on past experience may be influenced by changes in the practice of service companies and utilities. Furthermore looking to the increasing severity of weather conditions it may be necessary to demonstrate that the numbers used for transients leading to an SO event with safety injection temperatures at winter conditions are justifiable for each site.

Staff response: The temperature assumed for safety injection water in winter (40°F) is viewed as being a conservative bounding value that can be applied accurately or conservatively to all plants operating within the continental United States. Certainly occurrence of outside ambient temperatures in the deep South or in southern California of below 40°F can be considered a rare event.

- Item 2: As discussed already by other colleagues, the thermo-hydraulic calculation using RELAP-5 produce more or less mean temperature values in the downcomer at each time step. From the analysis results I have seen in the course of several safety assessments I conclude that nonuniform temperature distribution in the downcomer produce non symmetric loading conditions which have at least an impact on crack initiation of surface breaking flaws. The time of crack initiation and the orientation of flaws which would initiate are different from analysis results using purely symmetric cool down. It is difficult to judge for the reviewer if significant differences would result between nonuniform and uniform loading conditions for embedded flaws and cracks being extended to a considerable fraction of the vessel wall thickness. As it has been seen in the UPTF test the nonuniform condition caused by local mixing are not stable in space so model assumptions using the most pronounced nonuniform temperatures may be overly conservative for flaw locations beneath the surface. The validity of the arguments that a 2D RELAP calculation is sufficiently lined out in the summary report as well as other references should be judged by other colleagues.

Staff response: See comments made in response to Catton on this topic, and see defense of the RELAP model in [***Bessette***].

Probabilistic Fracture Mechanics

Major changes in the PFM methodology compared to previous studies are lined out in detail in the summary report and respective references. The most important ones are the change in the flaw distribution, the inclusion of the warm-prestress effect (WPS), the RT_{NDT} bias correction, the lift of the truncated value of K_{IC} and the inclusion of the crack face pressure.

In general these advancements reflect the increased understanding mainly based on experimental insights. There are no general objections to use these advancements in a regulatory context although the following 4 comments are directed to the conditions of application and the hardening of the justification.

Comment 1: Flaw Model

In my previous statement I made the following comment: "*Considerable effort has gone to the development of a more realistic flaw model by enlarging the experimental data sources. With the material available it is difficult to judge to what extent the sample material is representative for the whole set of vessels where the revised PTS Rule would be applicable. The reviewer is not familiar enough with the fabrication practice in the 60ies and 70ies as well as the differences in practice between the different manufacturers. To my knowledge the ultrasonic inspections during manufacturing in the 60ies and 70ies were largely voluntarily and not required by the code at that time. The in-service inspections following ASME XI are basically addressing welds. Concurrent with previous discussions (SECY/82/465) the reviewer would assume that a revised PTS Rule would also address the requirements on ISI and NDE qualification."* It is the view of the reviewer that a flaw model as outlined should only be used under conditions such as:

- Applicability check of the flaw density distribution for the pressure vessel under consideration including similarity check of fabrication practice.
- Applicability check of the flaw density distribution supported by non-destructive testing results for the near core region for weld and base material either using existing inspecting records or establishing a new finger print. In case only embedded flaws are used in the fracture mechanics analysis the necessary reliability of NDE to rule out surface breaking flaws may not be achievable by applying only ultrasonic methods, e. g. looking to one of the most recent exercises (NESC-1).
- It is the understanding of the reviewer that the flaw model is basically addressing remaining manufacturing defects. Although the operating experience with PWR vessels is judged to be favorable by the technical community it has to be remembered that the inspection of the cladding is not required by the ASME XI and being applied only in a few countries. Therefore the present view of the technical community may not be adequately based on inspection records.

It is the understanding of the reviewer that the flaw density distribution and material property distributions are used as independent variables. To my knowledge this is common practice but may not reflect the real situation for all kind of defects. From the experience of the past we have seen that crack like defects are governed to some extent by unfavorable material properties at certain locations. The reviewer admits there is no reliable data base to establish a correlation factor but still the sensitivity may be addressed in a parametric study assuming certain correlation factors.

The staff replies to this comment in Appendices B and C of NUREG 1806 is not sufficiently clear in my view. My interpretation of the staff response is that the flaw distribution can be used for PWR's in general without a specific check of plant records as lined out in my comment.

With the limited amount of material investigated and without detailed investigations of manufacturing records I do not share the views of the staff and Dr. Simonen that the flaw distribution can be applied in general to PWR's and judged to be conservative in nature.

Specific consideration should be given to documented repairs in base-, weld- and clad metal and their orientation with respect to location of high embrittlement.

<u>Staff response</u>: The staff maintains its position that the flaw distribution used in the FAVOR calculations includes sufficient conservatisms that it can be applied to the analysis of any PWR. Nonetheless, reviewer Schultz's concern on this point is noted, and we feel that his recommendation that use of the new screening limits should be tied to some in-service inspection (ISI) requirement is a prudent measure.

The basis assumption of no subcritical crack growth as lined out in chapter 3.3.3.2 reflects current thinking regarding environmental mechanisms and fatigue, but assumes regular conditions of water chemistry and cladding. Cracking of cladding have been observed at a number of steam generator primary side chambers and investigations of ripple loading and unusual quantities of chloride (which could be introduced in a maintenance action) could lead to corrosive driven crack extension (see /1/, /2/). This means that the basic assumption of non-surface breaking flaws and no operation induced crack and crack extension (Table 8.3 summary report) is bounded to well controlled conditions of operation and maintenance as well as in-service inspection of the cladding. Otherwise it may be difficult to justify the credibility of the flaw distribution.

> Staff response: It is the opinion of the staff that, even allowing for the possibility of chemical upsets, our fundamental assumption of no subcritical crack growth in the stainless steel cladding detailed in Section 3.3.3.2 (and, therefore, no surface-breaking flaws in the analysis) remains appropriate because chemical upsets will be of limited duration. Even in poor quality water (i.e., high oxygen) environments, Ruther et al. report an upper bound crack growth rate of $\approx 10^{-5}$ mm/s ($\approx 4\text{x}10^{-7}$ in/s) [Ruther 84]. The amount of crack extension that could occur during a chemical upset would is therefore quite limited, certainly not sufficient to compromise the integrity of the clad layer. It may also be added that stress corrosion cracking of the ferritic pressure vessel steel is not credible, even under upset conditions, because of the corrosion barrier provided by the stainless steel cladding.

The further question I raised in my previous comment (page B-58) regarding the treatment of flaw density and material property distribution is still valid. If no valid data base is available it is not unusual to establish a correlation factor by an expert elicitation process. This may be something to be addressed in a future R & D project.

> Staff response: Should sufficient information become available in the future (by expert elicitation or other means) on which a credible relationship between the existence of cracks and materials properties could be based, the sensitivity study suggested by reviewer Schultz would be appropriate.

Comment 2: WPS

The inclusion of the WPS effect is representing state-of-the-art. I was not able to check on the criteria applied regarding the slope of the decreasing K-field which limits the application of WPS. I assume that this is chosen in compliance with the uncertainly in the load transients.

Comment 3: RT_{NDT} bias correction and lift of truncation of K_{IC}

I believe that Eric van Walle is more qualified than my person to judge on these issues.

Comment 4: Validation of crack arrest

I have no principle doubt regarding crack arrest. My previous comment (page B-60) was directed to the state of validation regarding the validity of the calculated crack extension especially in the case of multiple events of crack extension.

The staff answer and the contribution of ORNL in Appendix E show the difficulty. The ORNL speaks of reasonable good agreement between experiment and analysis. Reanalysis of more recent experiments are not performed. If I take for example Fig. 104.3 (d) of Appendix F I have some doubts that the "accuracy" of the calculational procedure implied by this diagram is validated.

> Staff response: In the face of the large aleatory uncertainties characteristic of both crack arrest and crack initiation fracture toughness, it is the staff's view that the degree of agreement between deterministic predictions of crack run, arrest, reinitiation, rearrest behavior and

individual experiments that reviewer Schultz seems to desire is unrealistic. Because of these uncertainties exact agreement between predicted and measured crack initiation times and crack arrest lengths would be merely fortuitous. Consequently, a judgment regarding the appropriateness of the crack arrest model must be premised on the soundness of the underlying theory (see Chapter 5 of [***EricksonKirk-PFM***]) and the supporting experimental evidence provided by a limited number of structural experiments (see Appendix A to [***EricksonKirk-PFM***]).

3. Conclusion

The work performed show clearly advancements compared to previous studies. It is well founded in most parts. My major comments are directed to the flaw distribution and connected requirements to the plant-specific applicability as well as some reservation concerning the level of validation of crack arrest.
It is difficult to judge if a possible impact of these arguments are covered by the margins still existing in the presented numerical results.

References:

/1/ H.-P. Seifert, S. Ritter and J. Hickling: Environmentally-Assisted Cracking of Low-Alloy RPV and Piping Steels under LWR Conditions. 11th Int. Conf. On Environmental Degradation of Materials in Nuclear Power Systems – water Reactors, NACE/TMS/ANS, Stevenson, Colorado, USA, August 2003.

/2/ S. Ritter and H.-P. Seifert: The Effect of Chloride and Sulphate Transients on the Environmentally-Assisted Cracking Behaviour of Low-Alloy RPV Steels under Simulated BWR Conditions. EOROCORR 2003, Budapest, Hungary, September/October 2003.

Final comments made by reviewer VanWalle

25/11/2004
Peer Review Evaluation of the Draft USNRC PTS-Rule PFM-part CO 90 03 1729.00
TEC.50/B032076/05/EvW
van Walle Eric (33 3000), Head of Department

1. Appreciation

USNRC has made a systematic effort to review the actual 10 CFR 50.61 rule, also called the PTS-rule. From the outcome and review of the main document NUREG-1806 11-2-04 and supporting documentation NUREG 1807 11-7-04 + Appendices, it may be concluded that:

1. The newly proposed PTS-methodology is worked out well and has a logical and acceptable pattern. The separation and relation between the three parts -- PRA, TH and PFM -- is elaborated in a systematic and consistent manner and is a justified approach. The methodology is very well established, explained and documented in NUREG-1806;
2. A major improvement from the former report comes from the inclusion of the sensitivity studies that were performed on all steps and that clearly show the robustness of the overall approach and demonstrate the applicability of the methodology to NPPs in general;
3. Within the PFM-part, NUREG-1807, major ideas are founded on recent evolutions in the fracture mechanics community. These ideas are included within a framework that is based on continuity with the information extracted from the existing surveillance practices of the nuclear power plants (NPP);
4. The use of existing information -- obtained in the frame of reactor pressure vessel (RPV) surveillance programs or extra research oriented projects -- to define the models used in the PFM-part of the new PTS-methodology has the advantage that the NPPs are not requested to collect additional data on their RPV material to use the approach. At the same time it limits the direct application of a number of well-established innovative approaches (like direct fracture toughness determination via the Master Curve) within the procedure. The consequence of this indirect methodology tends to increase the uncertainty on a number of parameters used in the methodology.
5. The models used within the PFM-part can be considered innovative and are at the same time realistic. They are mostly based on/ derived from existing accepted models. A few models are however new and are validated on qualified but limited data sets. In time these data sets will need to be extended to further qualify and validate the suggested models;
6. Although no big changes on the outcome appear, the alterations in some of the modeling aspects of the PFM-part (a.o. the upper shelf model) make the approach more consistent and acceptable from the physics viewpoint;
7. When applied, sensitivity studies on the use of the models deployed in the PFM part demonstrate an acceptable realism and conservatism of the approach. The sensitivity studies on concurrent, resembling-equivalent models also made the USNRC team to select the model that gave the most conservative outcome;
8. The uncertainty treatment within the PFM-part is based on a classification scheme: a parameter is, or epistemic, or aleatory. This separation is very well suited in a probabilistic approach methodology and defines the way the uncertainty is propagated thru the overall

procedure. The occurrence of a non-unique classification, or a 'mixed' parameter gives however difficulties in this scheme;

9. The reviewer still believes that uncertainties on specific correlations used – unless clear proof is given that they would be double counted -- should be accounted for in the methodology. An example is the relationship between ΔT_0 and ΔT_{41J};

Staff response: The staff believes that the information presented in Figures 4.35 and 4.40 of NUREG-1807 demonstrate that simulation of the correlation uncertainties (as suggested by reviewer van Walle) would represent a double-counting of uncertainties, which is inappropriate. This is a point of disagreement between the staff and reviewer vanWalle.

10. The inclusion of a crack arrest model and the WPS effect is highly appreciated and adds to the realism of the methodology. More understanding towards crack arrest (when does it arrest, multiple arrests) is however needed and should be part of the USNRC recommendations for future work;

11. The information on the flaw distribution in vessel structures is based on a limited data base and might need more justification before generalization for representatively of all NPPs can be accepted. In service inspection remains at all times during the NPP life an important measure to be taken.

Staff response: The staff maintains its position that the flaw distribution used in the FAVOR calculations includes sufficient conservatisms that it can be applied to the analysis of any PWR. Nonetheless, reviewer vanWalle's concern on this point is noted, and we feel that his recommendation that use of the new screening limits should be tied to some in-service inspection (ISI) requirement is a prudent measure.

2. Recommendations

The reviewer, mainly oriented towards the PFM part of the procedure, recommends that:

- The PFM procedure as implemented in FAVOR 04.1 shall be used in the overall approach of the PTS methodology;
- The models used in the PFM procedure shall be reviewed on a regular basis to include upcoming data sets that can contribute to further validation of the models or to the reduction of the overall uncertainties in the procedure;
- In time, the possibility should be created to use direct fracture toughness data within the FAVOR procedure;
- In order to realize this aim a recommendation or obligation to obtain fracture toughness data from existing reactor pressure vessel surveillance materials of NPPs should be issued. This could be seen as a token for the use of the new PTS-procedure;
- As crack arrest is explicitly included in the PFM procedure, in time more information on this phenomenon needs to be gathered to validate the modeling;
- The flaw distribution needs more attention and validation for generalization and the importance of in-service inspections for flaw detection should also be stressed after eventual acceptance of the new methodology.

3. Specific comments

A limited number of specific comments will be sent to Mark EricksonKirk. These comments all relate to some textual inconsistencies in NUREG-1807.

4. Conclusion

The effort that has been put in reviewing the 10 CFR 50.61 PTS Rule by USNRC is more than substantial: it uses a logical framework, gets a massive amount of input scenario's and data from actual plants, removes a number of conservatisms in the old procedure, uses modern concepts of fracture toughness methodologies as a basis for elaborating the PFM modeling, ...

The PFM part can be accepted within the overall procedure and a number of recommendations on improvements have been formulated to justify the use of the approach.

The total effort comes to a trustworthy procedure that should allow, from the PFM part, relaxation of the actual 10 CFR 50.61 PTS Rule. However, the reviewer believes that after relaxation, some stringent conditions should be put on the NPPs that use the new procedure: statistics to more underbuild the procedure can only be obtained by testing their materials.

Eric van Walle.

Final comments made by reviewer Catton

TO:	*Thomas E. Murley, Chair*
SUBJECT:	*PEER REVIEW OF THE PTS TECHNICAL BASIS*
DATE:	*27 November 2004*
COPIES TO:	*David Johnson*
	Kumar Rohatgi (BNL)
	Helmut Schulz
	Eric van Walle
	Shah Malik

1.0 Introduction

In this report, the method developed by the USNRC staff to address modification to the existing PTS Rule, 10 CFR 50.61, is reviewed. The comments that are herein focus primarily on the thermal-hydraulic aspects of the proposed method of evaluation. The estimates of reactor vessel failure probability demonstrate a great deal of conservatism in the existing rule and support relaxation of the regulations to reduce the excess conservatism.

There are three parts to the estimation of the probability of vessel fracture resulting from rapid cooling. Each part will be discussed in the order of their occurrence in the estimation of the probability and its uncertainty. It is not by accident that thermal-hydraulics is placed second in the series of three parts. Thermal hydraulics is the circulatory system of a nuclear power station and, in this case, is what connects a probabilistic or reliability analysis to the vessel wall where structural mechanics takes over.

There has been an effort to treat uncertainties and the authors of the many documents are to be congratulated for their efforts. What is missing, however, is treatment of the propagation of uncertainties through the three step process. It would have been very helpful to have selected one or two typical sequences (for example, sequence 60 for the Palisades plant) and propagate the uncertainties from initiation to predicted failure probability of the vessel. This part of the process is circumvented by binning and treating the thermal-hydraulics as deterministic. The explanations are mostly reasonable but not quantifiable. Whatever uncertainties there are in the probability of an event and its descriptors could well be amplified by the uncertainties in the thermal-hydraulics.

It was disappointing to find that the thermal-hydraulics was assumed to be uncertainty free and to be told that the thermal-hydraulic calculations were best estimates. This may be the case for large-break LOCAs but is not the case for small breaks and other events with similar thermal-hydraulic behavior where our computational tools are less than adequate. Using arguments about relative uncertainty may be correct but adds uncertainty, and unease, in its own right.

Staff response: It was not assumed that the thermal-hydraulic analysis was free of uncertainty. Rather, assessing the impact of thermal-hydraulic uncertainty must be done in a deterministic manner, since the entire time-history of a given transient must be modeled as a boundary condition to the fracture mechanics analysis.

It is indeed true that RELAP5 is a best-estimate thermal-hydraulic code that has been extensively qualified for realistic analysis of small-break LOCAs and transients. In fact, a dedicated assessment effort was performed specific to the current PTS analysis to quantify the code to predict downcomer temperature, pressure and heat transfer coefficient [***Fletcher***].

The amount of material sent for review was overwhelming and due to time constraints was not all covered. This report is written with the knowledge that answers to many of the comments are in the next report. Often NUREG-1806 did not make reference to reports that could be found. It is clearly a report in progress. Some figure references are incorrect and arguments being given could not be followed. Nonetheless, it is hoped that the remarks will be found useful.

2.0 Initiating Event and Progression Probability

The treatment of events that could lead to a serious PTS event was one of the most thorough I have seen although it is difficult to assess completeness without spending a great deal more time. In this I will yield to my colleagues who are more involved in PRA. One of the outcomes of the evaluation, however, was a surprise. Arguments were given as to why the LOCA is more important than the events previously thought to dominate initiators. The basis for this surprising outcome is the role of the "subcool meter". This was an issue when 10 CFR 50.61 was developed around 1980 as there was always the question of whether one was faced with a SBLOCA or a PTS event.

> Staff response: The change in outlook is attributable to the fact that large LOCAs were analyzed in the current study and not in the earlier IPTS study. The operator procedures to respond to LOCAs and transients have changed significantly since the IPTS study to become more symptom-oriented.
>
> The only time subcooling entered into the analysis was with respect to tripping the reactor coolant pumps and determining when HPI throttling was permissible for repressurization scenarios (stuck-open pressurizer SRV that recloses).

The purpose of the PRA event sequence analysis is to obtain boundary and initial conditions for the thermal-hydraulic calculations. The boundary conditions and initial conditions are used to initiate computations using RELAP5/Mod 3.2.2gamma and the resulting time dependent downcomer temperatures and heat transfer coefficients are forwarded to FAVOR for calculation of the vessel conditional failure probability. It was argued, without quantification, that the variations in the PRA based boundary and initial conditions are far more important than the uncertainties in the thermal-hydraulic models in RELAP5/Mod3.2.2gamma. It would help this reviewer a great deal to have this assumption be quantified. Much of what follows could well be unimportant.

> Staff response: The accuracy and uncertainty in RELAP5 to predict downcomer temperature, pressure, and heat transfer coefficient was determined through comparison of the code with integral system experimental data. The uncertainties in these three parameters determined through assessment was shown to be small compared the variations in these parameters that occur from different sequences within a PRA bin [***Bessette***].

Again, the PRA process is one of the most detailed I have seen and, in particular, the iterations between the plant personnel and the NRC appears to have been very productive. The evaluation of 140,000 different possibilities by PRA is staggering.

3.0 Thermal Hydraulics

The thermal-hydraulic evaluation was done using RELAP5 3.2.2Gamma. The study was initiated by forming a PIRT group to delineate what was important and what was not. The results of the PIRT effort can be found in NUREG/CR-5452 dtd Feb 1999. The report documents the PIRT effort where the goal was to determine the thermal-hydraulic phenomena that have an impact on the temperature, pressure and heat transfer coefficient histories in the downcomer region around the core active length. This was done for a main steam line break from hot standby conditions, overfeeding of all steam generators from full power, and 2 in. breaks in both the cold and hot legs.

A number of phenomena were chosen by a panel and then ranked by the panel. The heat transfer at the wall, see Section 5.5.5, is described and the heat transfer coefficient is noted to be essential if the cool down rate of the wall is to be determined. A total of twenty two phenomena were ranked and the "heat transfer coefficient" fell below the cutoff. The basis for eliminating the heat transfer coefficient was the argument that "- - - the wall heat transfer coefficient is usually conduction limited in the vessel wall and the fluid velocities are relatively low with respect to the rated flows." That this is not the case has been known since the first PTS resolution.

Both the wall flux and the time rate of change of the interface temperature are strong functions of the time history and magnitude of the heat transfer coefficient and fluid temperature. Studies in the earlier visit to the PTS issue showed that the values of the heat transfer coefficient calculated using correlations now in the codes like RELAP5 fell midway between the wall conduction limit (very high heat transfer coefficient) and the convective limit (low heat transfer coefficient). The relationship between the heat transfer coefficient and failure probability for the base case was very steep.

Staff response: Evaluation of the Biot number shows that the heat transfer (heat flux) is indeed conduction-controlled. This has been shown many times by many different investigators. Iterative solution of the coupled conduction-convection equations demonstrated this as well. The incorporation of a model for free convection in RELAP5 provides a "floor" for heat transfer. Free convection gives higher values for heat transfer coefficient than forced convection at low velocities in the bulk fluid flow. Such is generally the case during conditions of loop flow stagnation, where heat transfer coefficients generally fall in the range ~1500 to 3500 W/m2-C.

That being said, the characteristic length term to be used in the Biot number analysis is most important. This length term must be selected considering the physical processes that control the fracture analyses. Using a different, smaller characteristic length indicates that the heat transfer coefficient has a greater importance than has been considered in past studies. Consequently, a number of sensitivity studies were performed to determine the influence of heat transfer coefficient, and these have been reported (NUREG-1809). The effect is most evident at conditions of loop flow stagnation and rapid cooldown of the reactor coolant system, characteristic of medium and large LOCAs.

We regard the subject of heat transfer (h) as one of but a myriad of sources of uncertainty that has been treated in the overall analyses. There is nothing unique or distinctive about it, particularly with respect to the other two important thermal parameters of temperature (T) and pressure. In fact, h should be considered together with T as part of heat flux uncertainty.

The twenty two phenomena were reduced to seven for consideration in determining the bounds or uncertainty in the thermal-hydraulic analysis. The heat transfer coefficient was ranked tenth leaving it out of further consideration. It was then argued that previous work and RELAP5/MOD3 development assessments had shown it to be adequate to predict these phenomena. This is not the case. RELAP5 may be adequate for predicting the phenomena under large break conditions but not for a small break. There are a number of problems that arise when the downcomer thermal-hydraulic behavior must be predicted and there is countercurrent flow in the cold leg. There is a missing part to the story we heard. The PIRT should have led to statements about processes that are important. Comparisons of code predictions with data from facilities that have been shown (by a scaling analysis) to be relevant then lead to knowledge of the uncertainties in the predictions. The uncertainty in a code prediction needs to be evaluated by itself. Application to a plant with appropriate consideration of operational uncertainties then yields the value and uncertainty of the final result.

Staff response: No basis is provided for the extreme statement that RELAP5 is inadequate for small-break LOCAs. To the contrary, RELAP5 has been extensively assessed against a large database of small-break LOCA experiments in a number of different integral system test facilities. Over the past ~25 years, the code has been applied extensively to many different small-break analyses. Recently, for example, the staff conducted an extensive effort to assess RELAP5 for analysis of AP600, and concluded that RELAP5 was applicable for such analyses. Nonetheless, PTS-specific assessment was performed to evaluate the performance of RELAP5 for predicting downcomer conditions for a spectrum of PTS-significant scenarios [***Fletcher***].

During a presentation of the thermal-hydraulic results, it was noted that a non-physical results led to putting in an artificial flow resistance to make flow go the way they thought it should. The presenter did not know what other effects this might have had on the final results. It was further noted that the downcomer fluid temperature forwarded for use in FAVOR was a spatial average. It seems to this reviewer that averaging the temperature in the downcomer rather than giving the PFM analyst the lowest value is nonconservative. The argument given for using the average was that the temperature differences were small and it does not matter. Such assumptions would be more acceptable if they were supported by more than judgment. As a result of questions raised at the last meeting with the staff, a RELAP5 sensitivity study was initiated and carried out. It is discussed in the next section.

Staff response: Whether calculated flows are "numerical" or physical was investigated as part of the PTS reevaluation. Specific assessments were performed, as well as sensitivity studies. As noted in the above response, a large body of experimental data was examined from a number of different experimental facilities. The data show the downcomer to be well-mixed in both the axial and circumferential directions. Plumes were found to be either weak or nonexistent. The maximum temperature variations observed in the experiments were comparable to the standard deviation of RELAP5 for predicting nominal temperature (σ ~10°C).

Use of average values for downcomer temperature and heat transfer rather than minimum and maximum values is appropriate. The comparison of RELAP5 with the experimental data for temperature was done on this basis. At any rate, there is little difference between the average value of temperature and the minimum value of temperature based on the well-mixed nature of the downcomer [***Bessette***].

3.1 RELAP5 Sensitivity Studies for PTS

A number of issues were raised during the course of the Peer Review and nine of them were addressed and reported by Bessette (RELAP5 Sensitivity Studies for PTS, October 2004). The report was reviewed and the results of the review are reported here. Bessette used the Palisades transients given in Table 1 to carry out his study. Seven of the nine issues are discussed in what follows.

3.1.1 Cooldown Rate

This was studied by approximating the downcomer fluid temperature by an exponentially decaying temperature and varying the decay rate and heat transfer coefficient. The downcomer temperature is given by

$$T_{dc}(t) = T_{ECC} + (T_0 - T_{ECC})e^{-bt}$$

The cooldown rates examined are shown in Fig. 1, taken from the thermal-hydraulic sensitivity study by Bessette. The RELAP5 result was used as a basis and higher and lower values of the decay rate were obtained by using higher and lower values of *b*. This is a reasonable approach if the initial value calculated by RELAP5 is appropriate. The ranges of values of *b* chosen, however, yield less than a plus or minus 20% variation in the total downcomer temperature change. The ROSA tests show larger variations than this although it is not clear whether or not the average decay rate is higher or lower. Further, it is not clear how the figure was used; e.g., the given temperature time behavior is used for all the transients. This temperature range needs to be justified and how it is used needs to be explained.

Staff response: The sensitivity study was intended to be illustrative and not comprehensive. It is true that faster cooldown rates do occur. The study looked at the effect of both cooldown rate and heat transfer coefficient, and concluded that of the two, the cooldown rate had a greater effect. The same set of temperature curves was used for the three studies of the effect of heat transfer coefficient (1.0 x h; 0.7 x h; 1.56 x h).

The study does illustrate that PTS scenarios can be approximated by simple exponential temperature decay, describing an ideally mixing (back-mixed, mixing cup) situation. When the exponential equation is fitted to a RELAP5 calculation, sensitivity studies can be performed easily.

The initial temperature for the calculations is the initial temperature of the downcomer and the reactor vessel during normal operating conditions. It is not dependent in any way on RELAP5 prediction. The transient selected as the basis for the sensitivity studies was a risk-significant sequence in Palisades.

Table 1, from Bessette (2004)

Palisades Transients Used as Basis for Sensitivity Studies

Palisades Case	Transient Mean	Initiating frequency
19	1 SG ADV stuck-open	2.3 E-3
40	16-in. HL LOCA	3.2 E-5
52	1 SG ADV stuck-open	6.4 E-4
54	MSLB	4.3 E-6
55	2 SG ADVs stuck-open	2.7 E-4
58	4-in. CL LOCA winter	2.7 E-4
59	4-in. CL LOCA summer	2.1 E-4
60	2-in. HL LOCA winter	2.1 E-4
62	8-in. CL LOCA winter	7.1 E-6
63	5.7 in. CL LOCA winter	6.1 E-6
64	4-in. HL LOCA summer	7.1 E-6
65	SRV recloses @ 6000s	1.2 E-4

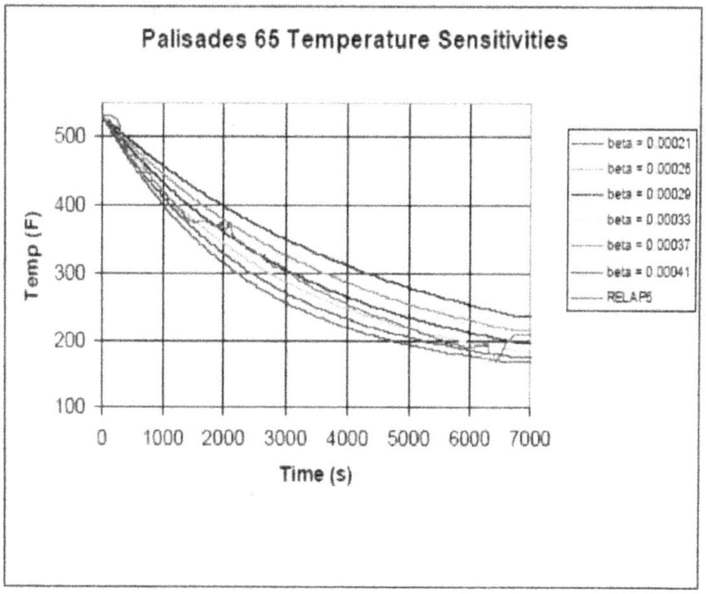

Figure 1
Cooldown Transients for Palisades

How it was decided that the cooldown rate predicted by RELAP5 is the appropriate starting place is not clear. It seems that this would be a good place to use the results from the OSU testing as it was based on scaled studies. Doing so would give a "code independent" evaluation as well as an opportunity to further validate the code. An initial step might be to incorporate some of the OSU scaling study results (NUREG/CR-6731) into Figure 1.

The version of RELAP5 used to generate the comparisons given in Table 2 (Sensitivity Analysis for Exponential -- - -") on page 5 of the Bessette report is not given. Further, the discussion indicates that the maximum decay rate used led to a 2.6 fold increase in CPF with a heat transfer coefficient multiplier of 1.0 and 3.4 with a multiplier of 1.56. This is somewhat confusing given the large numbers seen in table 5 on page 8 of the Bessette report.

Staff response: RELAP5/MOD3.2.2γ was used for these analyses, the same version that was used for all the PTS calculations.

Some comparisons of APEX and Palisades are given in Chapter 5 of "Scaling Analysis for the OSU APEX-CE Integral System Test Facility", NUREG/CR-6731 by Reyes. The calculations were done using REMIX and the scaled results compare reasonably well. Given the inapplicability of RELAP5 to small break or similar events, use should have been made of REMIX as it has undergone a great deal more scrutiny for such applications.

Staff response: REMIX calculations were performed and are documented in NUREG-1809. The results from these calculations are consistent with the experimental data and the overall conclusions regarding the extent of mixing in the downcomer. REMIX has been assessed against the separate effects mixing experiments, as well as the same UPTF Test 1 used for RELAP5 assessment. The results show that REMIX calculates the separate effects experiments well, with some conservatism in its prediction of plume strength.

However, from the amount of mixing observed in integral system test data, we find REMIX to have greater conservatism that indicated by its comparisons to separate effects tests. This is due to several limitations in REMIX and the separate effects data for which it was assessed:

- Modeling of the downcomer mixing as the decay of a single free plume is not appropriate.
- Most separate effects tests, with exception of IVO, did not investigate multiple plume interactions and plume merging.
- Heat transfer across the core barrel promoted by core decay heat is not included in REMIX or the separate effects experiments as a driving force for mixing.
- Heat transfer from vessel wall and other structures was not present in experiments where density effects are salt instead of thermal. The effect of heat transfer on mixing is not included in REMIX or separate effects experiments. Even thermal tests such as Creare had less that prototypic heat flux.
- In-vessel natural circulation driven by decay heat and ECC injection cooling flow is not included in REMIX or separate effect experiments. The flow circuit is up through the core, through the upper-plenum/downcomer the bypass, and down the downcomer (Theofanous found it necessary to model this effect in UPTF to get reasonable agreement between REMIX and the data)
- System flows promoted by break flow and depressurization is not included in REMIX or separate effects experiments.
- Complete annular downcomer compared to 90 degree unwrapped sector is not included in REMIX or separate effects experiments.

3.1.2 Downcomer heat transfer coefficient

How the average CPF for the sequences studied is found is not given. As a result, the basis for arriving at a CPF that is 3.3 times the RELAP5 base case is somewhat mysterious. Different configurations of RELAP5 were used to calculate the CPF:

- Base case: RELAP5/MOD 3.2.2Gamma (used for all the PTS calculations).
- Variation 1: RELAP5/MOD 3.3 (latest version of RELAP5).
- Variation 2: RELAP5/MOD3.3 with Petukhov-Catton implemented.
- Variation 3: RELAP5/MOD3.3 with Petukhov-Catton, and an additional heat transfer multiplier of 0.7 applied to hdc.
- Variation 4: RELAP5/MODE3.3 with Petukhov-Catton, and an additional heat transfer multiplier of 1.3 applied to hdc.

Table 5, page 8, shows that the CPF for some sequences is significantly increased, see sequence P-60, relative to RELAP5 Mod 3.3. Further, Table 5, page 8, shows increases in CPF ranging up to a factor of 25. Nothing is said in the report about what the differences between the two versions are. One has to wonder what the differences are between Mod 3.2.2Gamma and Mod 3.3 that led to the change and if it is real, why Mod 3.3 wasn't immediately substituted for Mod 3.2.2Gamma in the PTS study.

When the heat transfer package is changed to include mixed-convection, the increase in CPF for sequence P-60 is a factor of 25. There are other sequences that are increased by factors seven to twelve. If the heat transfer is increased 30% using Mod 3.3, the CPF is only marginally increased except for sequence P-40 where there is a twelve fold increase. The conduction limit has probably been reached and the result is no surprise. These results are indeed strange and deserve some attention. There is clearly a great deal of uncertainty in the thermal-hydraulic calculations.

Staff response: Mixed-convection is not considered relevant to the downcomer because of the significant enhancement in natural circulation flows in this region.

The above calculations were done as a sensitivity study with a developmental version of RELAP5. They were performed in response to a peer review request to explore the issue of mixed-convection. Because of the time constraint, the implementation of the heat transfer models, although reviewed, was not assessed at the time, or subsequently. More recent review of the results from cases 60, 40, 62 and 44 has raised doubts about the implementation of the heat transfer models because of the presence unphysical results.

Review of experimental data from Creare, UPTF, and APEX indicates downcomer mass flow rates substantially higher than those for which mixed-convection plays a significant role. For example, Creare reported a heat transfer enhancement attributable to mixed-convection of only ~5%. The characteristic velocities in the downcomer under flow stagnation conditions are 1 to 4-ft/s. See [***Bessette***].

3.1.3 Downcomer nodalization

It is argued that the 2-dimensional version of RELAP5 is conservative because the ratio of overall values of the CPF2D/CPF1D=1.5. Values of the ratio reach dizzying magnitudes under some circumstances (a ratio as high as 2.8 E7 for a 5.7 in. CL LOCA winter). With results like this, one can only conclude that there is something deeply malignant in RELAP5 or the writer forgot to tell us something. That the uncertainty in the thermal-hydraulic calculations is large should not be a surprise when using RELAP5 under conditions where it is inappropriate and has a "fix" that is known to be physically incorrect.

Staff response: We take issue with these extreme statements.

The results were explained in the sensitivity analysis. It is no surprise that the warmer the downcomer, the lower the CPF. Cold leg break flow patterns may differ significantly from hot side breaks. The favored flow path for a cold leg break is towards the broken cold leg. At a minimum, the ECC injected into the broken cold leg is bypassed out the break. Depending on size of the break and the time during the transient, some of the ECC injection into the intact cold legs may be bypassed as well. Indeed, the core and downcomer may experience flow reversal, so that water from the hot side of the reactor coolant system flows down through the core, through the lower plenum, and upwards through the downcomer towards the break.

Additionally, downcomer flows may be downwards near the intact cold legs, and upwards near the broken cold leg. Such flow patterns were commonly observed in large cold leg break experiments in UPTF. Naturally, use of a 1D downcomer precludes two dimensional flows. A 1D nodalization results in warmer downcomer temperatures. As a result, the 1D downcomer nodalization produced an average CPF that was more than a factor of 1000 lower than for the 2D downcomer.

Cold leg break LOCAs should not analyzed using a one-dimensional downcomer nodalization. Such a formulation means that the cold ECC injected flow is bypassed to the break through the upper downcomer instead of being allowed to flow to the lower downcomer. Downcomer temperatures, therefore, remain significantly warmer.

There is a long history associated with the desire to create a 2D downcomer for RELAP. The 2D downcomer was first used to evaluate a UHI plant (McQuire) in the late seventies. Values of velocities in the downcomer were known to be incorrect and very high but the overall result was what was expected, namely the UHI improved the LOCA result. Marshal Berman at SNL did the study for NRC. A brief study by ACRS consultants further confirmed that the use of a piping network to simulate 2 dimensional flow was physically incorrect and that by adjusting node-to-node azimuthal distances, one could get various results. The conclusions reached by Berman and by the ACRS consultants were that one should not use the piping network and if 2 dimensional behavior needed to be evaluated, a 2 dimensional code like TRAC or some other CFD tool should be used.

Staff response: The analyses referred to were performed by Sandia National Laboratory (SNL) as technical assistance to NRR. The investigations included analysis of upper head injection (e.g., McGuire), hot leg breaks and cold leg breaks. We reviewed work performed by SNL in the regard, including SNL reports NUREG/CR-0940, NUREG/CR-1364, NUREG/CR-1470, and NUREG/CR-1841. In addition we contacted remaining SNL and (former) NRR staff who performed the analyses. The analyses were performed at the time using RELAP4/MOD5 (as well as TRAC).

SNL performed several nodalization studies, and concluded that multi-channel was required in the core and the downcomer to obtain physically reasonable results, which was hopeless using a one-dimensional model. Therefore, one-dimensional nodalization should not be used in the downcomer and core in these circumstances. They also observed that numerical flow could occur (similar to the current codes), but that adequate modeling of form losses could prevent unphysical flows.

If the reference of the ACRS consultant were identified, we would be happy to review it. As it is, SNL reached the same conclusions 25 years ago that we repeated most recently.

3.1.4 Downcomer momentum flux

It is stated that "The 2D representation was employed because it provides for a better representation of the physical flow conditions. The additional degree of freedom renders loop flows more stable. It also allows a better representation of cold leg breaks, as was described above." It is correctly stated that the problem is with the cross flow terms in the momentum equation, but to argue that the 2D representation has any meaning is without basis. Several arguments are given as to why neglect of the momentum flux terms is not important. The least satisfactory is the statement "- - - thermal-hydraulic system codes do not, in general, conserve momentum, so the absence of the terms does not represent an important additional limitation." A caution about the appearance of numerically driven flows is given. It has been known since the seventies that these velocities can be quite high. High velocities do two things; 1) lead to higher heat transfer coefficients (increases CPF), and 2) better circumferential mixing of the fluid in the downcomer (reduces CPF). Comparing two versions of the code (one inappropriate and one physically incorrect) cannot lead to anything conclusive.

The table presented by Bessette, Table 7, page 11, demonstrates that the ratio of the CPF for the "momentum on" over the CPF for "momentum off" range all over the place. If momentum flux were relatively unimportant, the ratio would have been near unity. Instead it reaches a high of nearly 8,000 and a low of 1/10,000. This does not appear to be relatively unimportant to this reviewer. Granted, explanations are given for the observed results. The problem is, one must appeal to wide ranging arguments to gain comfort from what is presented.

> Staff response: Only two transients showed a substantial effect. These two transients were carefully evaluated. In reviewing case 156, a large-break LOCA, we noticed that the downcomer flows were excessive. While our review showed the problem to be limited to this one case, this one transient led us to repeat the calculation of the entire set of 75 Oconee PTS scenarios to ensure that we knew the overall results.
>
> To repeat the explanation accompanying the sensitivity study:
>
> "Two transients showed a large change (O-156 and O-110). O-156 was a large break (16-in.) LOCA, for which the CPF was a factor of 1000 lower with momentum flux off. This transient exhibited excessive downcomer circulation when momentum flux was on, which increased the heat transfer.
>
> Oconee 110 was a 2-in. surge line break with HPI failure. After 900s, the operator opened the two steam dump valves to lower primary system pressure and initiate accumulator and low-pressure injection. The two sensitivity calculations compared very closely for this transient, with almost no noticeable differences. At approximately 1830s, however, the case with momentum flux off, the pressure decreased slightly more than the case with momentum flux on.
>
> The small difference in pressure occurred when RCS pressure was ~200 psi for both calculations. The small difference was enough, however, to allow substantial LPI injection in one case but not the other (LPI shutoff heat = 200 psi). The difference in LPI injection flows caused a significant difference in downcomer temperature, which caused the difference in CPF. The particular transient is an excellent example of divergent, nearly bifurcating behavior that can occur as a result of plant design features such as: relief valve set points, level control, pressure control, pump shutoff heads, accumulator pressure, and so on.

It is beyond this reviewers comprehension why a code like TRAC was not used. It supposedly had overcome the momentum conservation problems and could correctly simulate a 2D downcomer. If there was institutional reluctance to use TRAC, then any one of a number of CFD codes could have been used. At a minimum, the reader should be shown a comparison with appropriately scaled experimental data so that an independent conclusion about the uncertainty in thermal-hydraulic calculations can be reached.

> Staff response: Use of the TRAC (TRACE) code was evaluated. However, at the time the code was under development and we experienced significant difficulties with run time and lack of robustness. In addition, the capability of TRAC (now TRACE) is in general similar to that of RELAP5 for predicting flows in the downcomer. While TRACE does include terms for cross-flow of momentum, the solution is not inherently improved. Both codes model the downcomer in two-dimensions using similar nodalization schemes. Neither code is able to model shear forces between fluids flowing at different velocities in parallel nodes.

3.1.5 Reactor vessel wall mesh size

A factor of ten reduction in node size reduced the CPF by one third. Several of the sequences (58, 59, and 63) demonstrated a sensitivity to the factor of ten increase in the number of nodes by increasing the CPF a factor of ten. All have relatively low CPFs but are sensitive nevertheless. One can only speculate what another factor of ten reduction in mesh size would do.

Staff response: We performed a nodalization sensitivity study to determine the appropriate mesh size to use, in which we varied the wall mesh in several steps between 8 and 80 nodes. This study showed convergence was reached at less than 80 nodes. From this study, we increased the nodalization from that used in the IPTS study (~8 nodes) to the 80 nodes used in the current study.

3.1.6 RELAP5-FAVOR boundary conditions time step

Matching the time step of FAVOR by using RELAP5 results at one second intervals caused very little difference in the final results.

3.1.7 Time averaging of RELAP5 output

RELAP5 time steps vary from 1 to 50 ms. CPFs for results averaged over 1 second were compared to CPFs for time averaged results and very little effect was noted.

3.1.8 Treatment of the cold leg flow

Although not called out specifically by Bessette, evaluating the impact of different temperature gradients is in part to try and delineate the impact of a one-dimensional single direction representation of countercurrent flow on mixing in the cold leg and flow into the downcomer, see Section 3.1.1. Countercurrent stratified flow occurs in the cold leg and cannot be analyzed by a code like RELAP5 without a great deal of uncertainty because RELAP5 is a one dimensional code. In the past mixing codes like REMIX, of which there are several, have been used to determine temperatures and flows at the cold leg inlet to the downcomer. REMIX is used by Reyes, see Chap 5 of OSU APEX-CE Integral System Test Facility, NUREG/CR-6731. REMIX is a simple multi-stage mixing code and is a reasonable tool although it needs some comparison with experimental data for corroboration. The mixing parameters, and their uncertainties, used in REMIX can be estimated. These can in turn can be used to estimate the uncertainty in temperature at various locations in the downcomer. This was promised in the PIRT report. Another alternative is CFD. The first study of the PTS issue using CFD was based on the COMMIX code.

Rather than attempting to address the actual problem, use is made of wide ranging arguments about relative unimportance of the results to the overall answer. The binning of the various calculations may well support this conclusion but it is not easily discerned that this is the case. Again, some quantification would have helped one to agree with the staff conclusions.

Staff response: We reviewed a large amount of data from several experimental facilities including LOFT, ROSA, Creare, UPTF, and APEX. The data show plumes to be either weak or nonexistent. The experimental data show substantial natural circulation flows in the downcomer that promote mixing, with mass flow rates 10 to 20 times the ECC injection flow rate. The mixing flows are consistent with the absence of significant plumes in the downcomer. COMMIX CFD calculations showed similar results [NUREG-1809].

We have assessed RELAP5 with experimental data for downcomer conditions including temperature, temperature distribution, and heat transfer. These assessment included comparisons with data for possible multidimensional effects. We have performed additional assessment of RELAP5 against integrated heat transfer experimental data from UPTF and

APEX-CE. This assessment shows good agreement between the base case RELAP5 code and the data.

3.2 NUREG 1806 Draft Dated 11-02-04, Section 6

As part of the analysis, key parameters and processes that affect the reactor vessel downcomer fluid temperature, primary system pressure and heat transfer coefficient on the inside of the vessel wall were defined. The Performance Ranking and Ranking Technique (PIRT) methodology was used to identify the most important processes that impact reactor system thermal-hydraulic response to a transient (see NUREG/CR-5452). It was not possible to find the document referenced in the list of references in the NUREG-1806. As a result, this reviewer is not sure whether or not the PIRT was revisited after earlier comments. It will be assumed that it was not.

Staff response: The reviewer has the necessary documents. We, in fact, revisited the PIRT on two occasions: (1) Reyes reconsidered the PIRT in his OSU testing program. NUREG/CR-6731, NUREG/CR-6856. (2) In the RELAP5 assessment report (NUREG/CR-6857), we also document our revision to the PTS PIRT.

Application of PIRT to the PTS issue yielded the following phenomena that should guide the code selection process:

- Break flow
- Primary system pressurization
- Natural circulation/flow stagnation
- Boiler-condensation mode and reflux condensation
- Mixing in the downcomer
- Condensation, mixing and stratification in the cold leg
- Integral system response

These parameters were selected because of their primary or secondary importance on downcomer conditions. The three phenomena judged to be of most importance to downcomer conditions during PTS events are:

- natural circulation/flow stagnation
- integral system response
- primary system pressurization

These phenomena were used to focus the RELAP5/MOD 3.2.2gamma assessment. Assessment was based on ability to predict the above phenomena. Mixing in the downcomer cannot be treated by any version of RELAP5. Further, although not noted in the PIRT effort, no version of RELAP5 can treat countercurrent stratified flow in a cold leg nor can condensation, mixing and stratification in the cold leg be dealt with. This oversight by the PIRT group remains with us.

Section 6.3.2, RELAP5 Numeric Issues, contains discussion of the occurrence of large azimuthal velocities when the downcomer is two-dimensional. The report states "The source of the circulation was traced to the application of the RELAP5 momentum flux model within downcomer regions that are represented using two-dimensional nodalization schemes (in the axial and azimuthal directions). The root cause of this problem in the RELAP5 code has not yet been uncovered, however it was found that deactivating momentum flux for the junctions within the downcomer region prevented these unphysical circulations. As a result, momentum flux was deactivated in the downcomer regions of the plant models used for the LOCA cases." In section 6.4 it is further stated that "The downcomer model used in each plant was revised to use a two-dimensional nodalization. This approach was used to capture the possible temperature variation in the downcomer due to the injection of cold ECCS water into each of the cold legs. Capturing this temperature variation in the downcomer is not possible with a one-dimensional downcomer nodalization. In the revised models, the downcomer is divided into six azimuthal regions for each plant. The reason for choosing six azimuthal regions is to match the geometry of the hot and cold legs around the circumference of the reactor vessel and so that water from each of the cold legs would flow into a separate downcomer node." Before such an approximation can be made, they should be quantified and shown to be valid. This could be done by order of magnitude comparisons of terms in the equations or comparison with data. This is particularly true when momentum flux is removed and the equations are an incorrect representation of the flow.

Staff response: We reviewed a large amount of data from several experimental facilities including LOFT, ROSA, Creare, UPTF, and APEX. The experimental data are consistent in showing the presence of a large degree of thermal stratification in the cold legs as a result of ECC injection of cold water into a system initially filled with hot water. The same data show downcomer plumes to be either weak or nonexistent. The experimental data show substantially enhanced natural circulation flows in the downcomer that promote mixing, with mass flow rates ~20 times the ECC injection flow rate. These large eddy mixing flows are consistent with the absence of significant plumes in the downcomer. The same behavior is seen at full-scale (UPFT), large scale full height (ROSA); large scale reduced height (LOFT), aspect ratio-scaled (APEX), and in separate effects tests (Creare). Similitude for fluid-fluid mixing and stratification is governed by Richardson Number scaling. The Reynolds number is influential as well. The different facilities were examined from this perspective to determine their applicability.

Any application of RELAP5 involves the solution of a thermodynamic control volume problem. The solution to such problems is governed by the initial conditions of the control volume, and the boundary flows across the control volume. Local phenomena may have importance as well, and these are identified through the PIRT process. The problem solution depends on both the model of the control volume as represented in the input deck, and the modeling of physical phenomena as represented by the code itself. Both influencing factors must be identified and ranked together; otherwise, effort may be wasted on unimportant parameters. A large number of sensitivity studies were performed in association with the PIRT to quantify the effects of the different boundary conditions and physical models in RELAP5.

RELAP5 was assessed against separate-effects experiments to evaluate its capabilities for predicting specific localized behavior that is relevant for PTS. These separate effects experiments included Marviken tests for assessing critical flow models, MIT Pressurizer facility tests for assessing steam condensation and RCS pressurization behavior, UPTF full-scale tests for assessing condensation and steam/water flow phenomena and Semiscale tests for assessing coolant loop natural circulation flow behavior. In spite of this impressive list, comparisons of important downcomer behavior like mixing or penalties associated with the lack of a countercurrent stratified flow model are given as overpredictions and underpredictions. This in itself is not enough, because as shown by Bessette, the key parameter is temperature decay rate.

Staff response: See previous response

The TH uncertainty characterization, Section 6.8.2.2,

In Bessette's sensitivity assessment, Mod 3.3 is used and was found to yield different results. The basis for not rerunning many of the sequences using Mod 3.2.2gamma is not given nor is any explanation for the differences. See comments on the Bessette sensitivity study. The momentum flux problem in RELAP5 is the non-physical nature of the momentum equations that results from the use of a piping network to represent 2 dimensional flow. The problem has been known to exist for 25 years. This is not new.

Staff response: See the first staff response in Section 3.1.4 of this letter.

4.0 Structural Mechanics

The advances in PFM since PTS was last addressed appear to be significant. When PTS was last addressed, the major uncertainties were in the PFM. It appears as if this has changed and that the values and uncertainties can be evaluated subject only to the uncertainties in the thermal-hydraulics. This area is left to my colleagues to discuss. It came as some surprise, however, that the LOCA is a dominant contributor to the PTS risk.

Staff response: As discussed in Section 8.5.2.5, medium- and large-break LOCAs never appeared as dominant transients in previous assessments of PTS because these transients were never analyzed. Our analysis shows (see Figure 8.44) that for highly embrittled vessels medium and large-break LOCAs contribute more than half of the total TWCF. However, at the more modest embrittlement levels characteristic of 40 to 60 years of operation stuck-open valves (primary side) are the dominant risk contributors

When experts in PFM were asked, they noted that they knew where more dangerous regions of the vessel wall could be found. Given this information, it is not clear why they were not some how weighted into the computational procedure. Everything is done in terms of averages. How can one do this when it is known that there are regions in the vessel wall that are more susceptible to thermal shock than others. Rohatgi used an analogy to describe this concern. He asked you to imagine a river that was, on average, 5 feet deep. The shore is shallow and there is a 20 foot deep trench along the center. You will surely drown if you try to walk across. How can this trench just be average under the guise of "PRA"?

Staff response: It is unclear to the staff the basis of reviewer Catton's opinion that "everything is done in terms of averages" in the PFM analysis, because this is not the case. Interested readers are referred to the various reports on the PFM analysis (see Section 13.1.3 for a complete list) as these describe in detail all of the location dependencies that are explicitly accounted for in our calculations. By way of summary, we can point out that location dependencies in the flaw population, in fluence, and in chemistry and fracture toughness properties are all simulated. Collectively, these simulated the spatial variability of the fracture resistance of the RPV steel.

Location dependencies in the flaw population

- Based on non-destructive and destructive evaluations of RPV vessel materials, we have determined that embedded weld flaws only occur on the weld fusion lines (that is, the interface between the deposited weld metal and the plate or forging that the weld joins). Consequently, embedded weld flaws are simulated ONLY to occur on the weld fusion lines. This means that the flaws associated with axial welds are ALWAYS oriented axially whereas the flaws associated with circumferential welds are ALWAYS oriented circumferentially.

- Based on non-destructive and destructive evaluations of RPV vessel materials, we have determined that embedded plate flaws can occur with equal likelihood at any location within the plate and with any orientation. Consequently, this is the way FAVOR simulates these embedded flaws.
- Our non-destructive and destructive evaluations have also determined that embedded weld flaws and embedded plate flaws have different size distributions (embedded weld flaws being generally larger). Consequently, FAVOR simulates these different size distributions ... preferentially placing the larger flaws along the weld fusion lines and placing smaller flaws preferentially in the plates.
- Finally, our non-destructive and destructive evaluations of RPV vessel materials have revealed that surface breaking flaws can arise as lack of inter-run fusion defects between the weld beads of the austenitic stainless steel cladding that is deposited on the inner diameter of the vessel. Consequently, the only surface flaws that FAVOR simulates are oriented circumferentially (because the cladding is laid down circumferentially) and are simulated to have a depth equal to the thickness of the cladding.

Location dependencies in fluence

- As illustrated in Figure 8.1 of NUREG-1806 the magnitude of the neutron fluence to which the vessel wall is subjected varies markedly in both the azimuthal and axial orientations due to the (respectively) the variable gap between the core and the ID and due to the finite axial length of the core. These variations cause proportional variations in the level of irradiation damage (i.e., toughness reduction) experienced by different locations on the vessel ID. Both the axial and azimuthal fluence variations illustrated in Figure 8.1 are simulated by FAVOR/
- Additionally, the level of neutron damage diminishes exponentially as you move through the vessel wall from the ID to the OD because the steel closer to the ID "soaks up" the neutrons ... thereby resulting in less damage to the RPV steel that lies further from the ID. This attenuation of neutron damage through the vessel thickness is simulated by FAVOR.

Location dependence of chemistry and toughness properties

- Each weld, plate or forging in the beltline of the vessel may have its own unique values for chemistry variables (Cu, Ni, P) and for toughness variables (RT_{NDT}). Thus the chemical composition (which controls the irradiation sensitivity) and the toughness of the vessel before irradiation is location dependent. FAVOR uses these mean values to center the distributions of Cu, Ni, P, and RT_{NDT} from which it samples to simulate the point to point material variability within the vessel.

5.0 Concluding Remarks

- It was argued, without quantification, that the variations in the PRA based boundary and initial conditions are far more important than the uncertainties in the thermal-hydraulic models in RELAP5/ Mod3.2.2Gamma. This assumption should be quantified. If this is done, much of what follows could well be unimportant.
- The PRA process is one of the most detailed I have seen and, in particular, the iterations between the plant personnel and the NRC appears to have been very productive. The evaluation of 140,000 different possibilities by PRA is staggering.
- Given the differences between predictions using Mod 3.2.2Gamma and Mod 3.3, a tenfold change in some CPFs, an explanation of what caused the differences is needed.

- Use of a two dimensional downcomer with or without momentum flux results in more questions than answers when no experimental confirmation is demonstrated to support the conclusions reached.
- Why after 25 years is there not a computational tool that can address two dimensional flows in the downcomer? Why after 25 years is there no computational tool capable of addressing countercurrent stratified flow in the cold legs? These are not new questions. There are many examples of what happens when such computational capabilities do not exist. The primary result is highly conservative and argumentative positions on safety issues leading to results in which little faith can be placed.
- Use of the "relative importance" of the computed results to argue that improper use of a computer code is acceptable without quantification does not give one confidence in the final result.
- When bin uncertainties are discussed, the uncertainties are the result of slightly different cases being put in the same bin. This has nothing to do with code uncertainty. Within one of these bins, the code uncertainty should be shown to complete the argument and if the code uncertainty is well within the bin uncertainty, then the code is good enough. This has yet to be done.
- The advances in PFM since PTS was last addressed appear to be significant.

Final comments made by reviewer Rohatgi

Review of Technical Basis for Revision of Pressurized Thermal Shock (PTS) Screening Limit in the PTS Rule (10 CFR 50.61): Summary Report; (NUREG-1806);
Thermal Hydraulic Aspects (Draft)
U.S. Rohatgi, BNL

NUREG 1806 is a well-written document describing the approach from the introduction of the problem, PRA, thermal-hydraulics to fracture mechanics. The approach is systematic and addresses full scope of PTS Rule changes. It captures the advancement in technology since 1982 in the area of PRA and fracture mechanics along with characterization of the flaws.

The thermal-hydraulic area has not advanced as much since 1982 as other areas. The RELAP5 code is still a one-dimensional code but it has become more robust. There are still some problems with mass and momentum conservation and condensation model.

The PRA provides the sequence of transients and their frequencies. These sequences are put in groups (bins) based similar behavior. A sequence is selected from each bin to represent that bin or class of transients. This selected transient is analyzed with a best estimate system code such as RELAP5. The analyses provides a history of temperature, pressure and heat transfer coefficient in the downcomer. This information is supplied to probabilistic fracture mechanics code (FAVOR). This code predicts the probability of through the wall crack. The thermal-hydraulic analyses are the bridge between PRA and fracture mechanics.

Here are my comments that are related to TH.

1. LOCAs have become more important than in 1982. Good technical basis is provided.
2. Statement (Page 6-103) such as there is no uncertainty in TH calculation should be removed. There is significant uncertainty in many two-phase flow models.
3. How is a representative sequence selected for the bin? Bin will include different sequences with different combination of downcomer fluid temperature and heat transfer coefficient. Last paragraph on page 6-106 is not clear.
4. How is the TH uncertainty factored in TWCF? This is not clear from the description.

<u>Staff response</u>: Section 3.2.2 of this report describes how uncertainties propagate though our analysis. Specifically, Section 3.2.2.4 describes what the uncertainties in TWCF represent. And points out that in each of the three technical modules (PRA, TH, and PFM) the uncertainties have been either "accounted for" in that they influenced the structure of the computational model, or they have been "numerically quantified" as part of that model. Thus, a description of what the uncertainties in the reported values of TWCF represent requires more than a strictly numerical answer. The numerical value of the TWCF is estimated by performing a matrix multiplication of the distribution of frequencies of each bin defined in the PRA analysis with the distribution of conditional probabilities of through-wall cracking estimated by the PFM analysis. However, these uncertainties (of bin frequency and of the conditional probability of through-wall cracking) and their quantifiable distributions arise as a direct consequence of the particular model we have used to calculate them. Thus, the structure of the model itself accounts for a number of uncertainties that have not been numerically quantified.

5. The flow phenomenon is three-dimensional in the downcomer. There may be mixed-convection due to circulation between core barrel and downcomer inner wall (radial direction), and flow in

azimuthal direction due to injections at discrete locations from the cold legs, and temperature distribution in the cold leg.

Staff response: See Comment #65.

The RELAP5 code has been modified to include Catton-Petukhov heat transfer coefficient to account for mixed-convection with flow circulation in axial-radial plane. It did show increase in TWCF. However, it is not clear if Catton-Petukhov correlation was used with two-dimensional downcomer model. There is also flow in azimuthal direction that is expected to be represented by two-dimensional model. The concern is that RELAP5 only approximates that flow. It is not clear why there is increase in azimuthal flow when momentum flux is included. However, it is conservative to include this term (see transient O-110). In general 2-D is conservative compare to 1-D, sometimes many orders of magnitude in TWCF. Please provide recommendation for TH calculations for RELAP5 and other best estimate codes.

Staff response: See Comment #65. We conclude that two-dimensional nodalization should be used in the downcomer. Applications of the code should include comparisons with experimental data relevant to the problem at hand.

6. Table 6.1 shows the important phenomena and boundary conditions. It will useful to show the tests that will cover these phenomena.

Staff response: The RELAP5 assessment carried out for PTS was based on the PIRT, as described in [***Bessette***].

7. RELAP5 validation is done with separate and integral effects tests. How is downcomer modeled in these tests, nodalization and momentum flux option? Also assessment results are indicated in terms of average over the transient. This will average out the large differences and even cancel out the difference in apposite direction. It will be better to provide the range of the difference (max and min).

Staff response: See NUREG/CR-6857. Also, assessment results for temperature and pressure reported as the combination of mean bias and standard deviation are more information and appropriate than maximum and minimum values, particularly because of the averaging nature of the vessel wall for short time fluctuations in temperature.

8. How is the uncertainty added? Report indicated that a linear addition will be sufficient and response surface approach may not be needed. However, is the linear additional method an addition of the magnitudes or algebraic values?

Staff response: The methodology for TH uncertainty treatment is detailed in Section 3.2.2 of this report, and in [***Bessette***]. The linear additivity referred to by this reviewer was discussed as part of the studies performed by the University of Maryland [***Chang***] that were used to guide bin subdivisions. Linear additivity of uncertainties did not enter into the mathematical expressions used to determine TWCF.

Conclusion

1. Methodology is an improvement over 1982 studies. It is logical and covers all the important aspects of the problem. The number of sequences is large and covers all type of transients. RELAP5 is a best estimate code and has become more robust and has been validated with large number of integral and separate effect tests.
2. The flow field in the downcomer is three-dimensional. There is need to show how RELAP5 can be used and what is the uncertainty? There is need to compare predicted flow field with the data (where available) and with CFD.

Staff response: In [***Bessette***], we show, on the basis of integral systems tests that represent the conditions in a PWR, that the degree to which the flow field is three-dimensional in the embrittled region near the vessel core that the three-dimensional effects can be ignored without loss of accuracy. In [***Bessette***], comparisons are made with CFD.

Final comments made by reviewer Johnson

Introduction

My review focused on the "PRA" aspects of the integrated analyses. Specifically, the review considered the selection and treatment of plant scenarios, the interface between the "PRA" and thermal-hydraulic portions of the analyses, the treatment of uncertainty and the generalization of the results and conclusions to the fleet of U.S. PWRs.

Identification of Plant Response Sequences

I will refer to the results of the "PRA" portion of the analyses as sets of plant response sequences. The analysts used the plant-specific PRAs performed by the utilities as bases for the identification of the plant response sequences. In some cases, these plant-specific PRAs required some augmentation to assure that sequences of potential PTS interest were included. This use of the plant-specific PRAs developed by the licensees – as well as interaction with the licensee technical staff – seems to me to be an example of utilizing the best available information. I believe that the approach taken resulted in more robust and complete sets of plant response sequences as compared to the alternative approach of using the more limited – although improving – plant-specific SPAR models. The cooperative use of the best available models should be encouraged.

The categories of sequences of potential PTS interest include stuck-open primary safety valves (or PORVs), medium and large LOCAs and secondary system upsets. The categories identified appear to be complete and logically identified.

The modeling of the different categories of sequences is consistent for the three plants and, moreover, conservative. The degree of conservatism, however, varies among the sequence categories. In particular, I believe that the treatment of inadvertent opening (or stuck-open) primary safety valves is particularly conservative. The primary source of conservatism is in how valve reclosure is modeled. I believe that reclosure, if it were to occur, would be most likely early in the transient when pressure in the primary circuit is still relatively high. I do not believe the assumed closure time bins of 50 and 100 minutes have an actuarial or engineering analysis basis. As noted in the report, the length of time the valve is open is one of the controlling factors of the PTS aspects of this category of sequences. Nonetheless, the model used in the analyses is certainly conservative. It is noted that this category of sequences is an important contributor to the estimated TWCF (e.g., more than 60% of the total estimated TWCF for Oconee) and that the conditional likelihood of a through-wall crack is relatively high. If the analysts agree with the relative degree of conservatism for the model of this category of sequence, then additional discussion of this point in the report may prove useful to potential future analysts performing plant-specific analyses.

Binning and Endstate Characterization

I was most interested in the linking of the "PRA" and "TH" portions of the analyses. The large number of sequences arising from the "PRA" portion of the analyses necessitates the use of binning sequences with similar characteristics so that a manageable number of TH calculations can be defined that represent the range of sequence conditions.

I was looking for a parallel between the binning performed in these analyses and that done linking the level 1 and level 2 portions of a PRA. In the latter, similar level 1 sequences are logically grouped such that the variation (as measured by the various level 2 model elements) among the individual sequences assigned to a single bin is small. While the binning done in the current analyses was done on an iterative ad hoc basis, it appears that the resulting interface is reasonable. It is noted that the interface allowed for insights to be gained with respect to the relative importance of the contributions of specific initial conditions defined in plant response to the estimated TWCF.

I do have a remaining question regarding the degree of conservatism surrounding the selection of the end state of the analysis. I recognize that previous work, including Regulatory Guide 1.154 equates TWCF with CDF. What is not clear to me is what insights the analysts could offer regarding the physical damage condition of a vessel that has experienced a through-wall crack. Do the fracture mechanics analysts expect the vessel to fail in such a way as to lose coolable geometry, or is this an assumption? Are there conditions that would make some failures (or through-wall cracks) more severe than others? If so, can these conditions be related back to specific scenario categories? This could be an place where the analysts could provide additional insights.

Accepting that equating TWCF with CDF is perhaps conservative to some unknown degree, the discussion of the selection of acceptance criteria is appreciated. The report recognizes that there is currently an incomplete understanding regarding the progression of an accident following a postulated PTS-induced vessel failure. Nonetheless, the report presents a well-reasoned framework that bounds the potential influence several complex issues such as the impact on the source term of fuel damage in an "air" environment.

Characterization of Uncertainty

In general, the treatment of uncertainty in the analyses is well thought out with a significant amount of effort invested in the effort. The focus of the uncertainty analysis seems to be on characterizing the frequency of the scenarios, with a lesser amount of effort invested in exploring and characterizing modeling uncertainty.

For example, the uncertainty associated with the PRA portion of the analysis focuses entirely on characterizing the frequency of the scenario rather than any uncertainty inherent in or introduced by the model. It should be noted that this approach is typical of the uncertainty treatment for level 1 PRAs for commercial nuclear power plants.

The TH analysis does list five phenomena that are claimed to represent the only significant sources of modeling uncertainty. While I am not a TH expert, it is not clear to me how these phenomena represent all the significant modeling uncertainty. For example, I do not understand how the limitations of the code to represent the spatial distribution of wall temperatures are reflected in the uncertainty analysis. After all, the code was designed to predict the temperature history of the fuel under certain conditions with minimal attention paid to the wall temperature. The TH analyses do contain several sensitivity cases that explore the potential impact of selected features. These analyses seem to all use the same basic model structure.

As I understand it, the insights gained in performing the TH sensitivity cases were used primarily in an iterative process to refine the PRA bins. New TH cases were defined to represent the newly partitioned PRA bins. The result is the "method of accounting for TH uncertainty does not quantify the uncertainties associated with each TH sequence, but rather it characterizes the uncertainties with each PRA bin." The report claims that "any errors caused by not explicitly accounting for TH parameter and modeling uncertainties associated with the TH sequence used to represent each PRA bin are not expected to influence the outcome of the analysis (i.e., the estimated values of TWCF)." I believe this approach blurs the distinction between sensitivity analyses and an uncertainty analysis. I am not convinced that the TH analyses contribute negligibly to the TWCF. I understand that the parameters investigated are shown to be small contributors, but I am not convinced that a full accounting is made for modeling uncertainty.

Generalization to Fleet of PWRs

A comprehensive framework is presented that seeks to permit a generalization of the (conservative) analyses of three specific plants to the fleet of U.S. PWRs. This framework is well reasoned and, as far as I can tell, complete. It offers a useful tool to determine if one of the existing cases can bound the impact of PTS for any specific plant.

Conclusion

The analysis team has accomplished an impressive task. This task was to apply – and in some cases extend – the current understanding of PRA, TH and PFM to revisit the basis for the PTS Rule. The analysis team has succeed in this task and has produced a body of work that that is, in my opinion, both reasonable and demonstratively conservative.

Appendix C – Flaw Distribution, Correspondence with Dr. Fredric Simonen of the Pacific Northwest National Laboratory

The following report details the flaw distribution adopted in FAVOR and used in this investigation.

Simonen 10-03	***F.A. Simonen, S.R. Doctor, G.J. Schuster, and P.G. Heasler, "A Generalized Procedure for Generating Flaw Related Inputs for the FAVOR Code," NUREG/CR-6817 Rev. 1, October 2003.***

This appendix includes the text of a letter sent to the primary author of this report, Dr. Fredric Simonen, and Dr. Simonen's response. The purpose of the letter was to clarify Dr. Simonen's views regarding the extent to which the flaw distributions reported in NUREG/CR-6817, Rev. 1 apply to operating PWRs *in general*.

Text of Letter Sent to Dr. Simonen

30th June 2004

MEMORANDUM

From: Mark EricksonKirk (mtk@nrc.gov)
To: Fred Simonen (fredric.simonen@pnl.gov)

cc: Debbie Jackson
Allen Hiser

Subj: NUREG/CR-6817, Rev. 1, "A Generalized Procedure for Generating Flaw-Related Inputs for the FAVOR Code," by F. A. Simonen, et al.

Motivated by comments received from both the external peer review panel we convened for the PTS project and from some members of the industry I have recently re-read the subject NUREG/CR report. For the PTS reevaluation effort it is important to know to what extent the flaw distributions reported in NUREG/CR-6817, Rev. 1 apply to operating PWRs *in general*. Neither the executive summary nor the conclusions of this report (which I have attached for your reference) speak to this issue. However, I did find the following statements in the body of the report that speak to the question of the general applicability of the flaw distribution:

On p. 5.9 (***emphasis*** added):

> The PRODIGAL model provided a systematic approach to relate flaw occurrence rates and size distributions to the parameters of welding processes that can vary from vessel-to-vessel. Application of the model showed the sensitivity of calculated flaw distributions to changes in the welding process conditions. ***Calculations with PRODIGAL and consideration of known differences in fabrication***

procedures used to manufacture U.S. vessels indicated that data from PVRUF and Shoreham can reasonably be applied to all vessels at U.S. plants.

On p. 6-2 (***emphasis*** added)

Use of Data Versus Models and Expert Elicitation - In developing flaw distributions, measured data were used to the maximum extent possible. The PRODIGAL flaw simulation model and results of the expert judgment elicitation were used only when the data were inadequate. In the case of seam welds, there was a relatively large amount of data, and the PRODIGAL model and expert elicitation were not used to quantify estimates of flaw densities and sizes. The PRODIGAL model did, however, suggest the normalization of flaw dimension by the dimensions of weld beads and the separation of data into subsets corresponding to small and large flaws (as defined by flaw depth dimensions relative to the weld bead dimensions). ***In addition, the expert elicitation and the PRODIGAL model helped to justify the application of data from the PVRUF and Shoreham vessels to the larger population of vessels at U.S. nuclear plants.***

The NUREG/CR also includes the following statement:

On p. 6-3 (***emphasis*** added)

Vessel-to-Vessel Variability - The PNNL examinations of vessel material focused on two vessels (PVRUF and Shoreham), with only limited examinations of material from other vessels (Hope Creek, River Bend, and Midland). The Shoreham flaws showed some clear differences from the PVRUF flaws with somewhat greater flaw densities and longer flaws (larger aspect ratios). However, there was no basis for relating these differences in flaw densities and sizes to other vessels. With only two examined vessels it was not possible to statistically characterize vessel-to-vessel differences such that the differences could be simulated as a random factor in Monte Carlo calculations. The decision was to develop separate procedures to generate flaw distributions for the PVRUF and Shoreham vessels. ***Following the conservative approach taken in other aspects of the PTS evaluations where data and/or knowledge is lacking, it was recommended that the Shoreham version of the flaw distribution be used in PTS calculations, which served to ensure conservatism in the predictions of vessel failure probabilities.***

The statements from p. 5-9 and 6-2 suggest that the view that the flaw distributions proposed in NUREG/CR-6817, Rev. 1 apply to operating PWRs *in general*. Conversely, the statement made on p. 6-3 seems to suggest that you and your co-authors view the flaw distributions as being *conservative*.

To help me respond to questions I have received regarding use of the flaw distributions presented in the NUREG/CR in the PTS reevaluation project it would be most helpful to me if you could respond to the following question:

What is the view of you and your co-authors? Do you view the flaw distributions published in NUREG/CR-6817, Rev. 1 as being applicable to PWRs in general, or do you view them as being a conservative representation of the flaw population in the fleet of operating PWRs.

I greatly appreciate your assistance with this matter.

Reply Received from Dr. Simonen

>>> "Simonen, Fredric A" <fredric.simonen@pnl.gov> 07/01/04 02:23PM >>>

Mark:

This is my response to the questions that you posed to me in the attached memo (June 30, 2004):

What is the view of you and your co-authors? Do you view the flaw distributions published in NUREG/CR-6817, Rev. 1 as being applicable to PWRs in general, or do you view them as being a conservative representation of the flaw population in the fleet of operating PWRs?

Your June 30, 2004 memo accurately reflects my views and those of my co-authors regarding the applicability of the flaw distributions in NUREG/CR-6817, Rev1 to PWRs in general as well as the conservative nature of the distributions.

In developing the flaw distribution methodology we were guided by Lee Abramson (statistician from NRC staff) in dealing with uncertainties. Because the PNNL flaw data were primarily from two vessels (PVRUF and Shoreham) a rigorous statistical treatment of vessel-to-vessel differences was not possible. The flaw model was therefore developed with separate treatments for the two vessels, along with a recommendation to use the more conservative treatment based on the Shoreham vessel when addressing other vessels. Other conservatisms can be introduced as appropriate in application of the flaw model to address uncertainties in knowledge regarding of a specific vessel. One example of such uncertainties would be the amount of repair welding in a particular vessel.

Fredric A. Simonen
Laboratory Fellow
Pacific Northwest National Laboratory
P.O. Box 999
2400 Stevens Drive
Richland, Washington 99352 USA
phone 509-375-2087
fax 509-375-6497
eMail fredric.simonen@pnl.gov

<<d jackson memo 30 jun 04.doc>>

Appendix D –Comparison of Plant-Specific Reference Temperature Values

PWR Plant Name	RT_{PTS} at EOL from RVID [°F]	RT at 32 EFPY (EOL) [°F]			RT at 48 EFPY (EOLE) [°F]			TWCF Estimated Using Eq. 11-2	
		RT_{MAX-AW}	RT_{MAX-PL}	RT_{MAX-CW}	RT_{MAX-AW}	RT_{MAX-PL}	RT_{MAX-CW}	32 EFPY	48 EFPY
ARKANSAS NUCLEAR 1	237	118	93	173	127	102	184	7.7E-12	1.3E-11
ARKANSAS NUCLEAR 2	123	105	105	105	117	117	117	3.6E-12	7.3E-12
BEAVER VALLEY 1	268	187	228	219	198	243	236	7.2E-10	1.5E-09
BEAVER VALLEY 2	153	85	104	104	88	116	116	1.3E-12	1.8E-12
BRAIDWOOD 1	85	0	33	82	0	36	88	1.3E-14	1.4E-14
BRAIDWOOD 2	70	0	49	78	0	51	83	2.1E-14	2.3E-14
BYRON 1	110	0	78	78	0	84	90	8.1E-14	1.1E-13
BYRON 2	103	0	38	105	0	42	121	1.6E-14	2.2E-14
CALLAWAY 1	115	81	88	88	85	93	93	9.0E-13	1.2E-12
CALVERT CLIFFS 1	253	193	156	156	204	171	171	6.8E-10	1.3E-09
CALVERT CLIFFS 2	198	167	167	167	179	179	179	1.5E-10	3.0E-10
CATAWBA 1	58	0	44	18	0	48	22	1.8E-14	2.0E-14
CATAWBA 2	128	93	93	93	99	99	99	1.7E-12	2.6E-12
COMANCHE PEAK 1	100	67	67	67	75	75	75	3.7E-13	6.0E-13
COMANCHE PEAK 2	92	39	39	39	43	43	43	6.9E-14	8.8E-14
COOK 1	215	153	162	202	166	173	217	6.7E-11	1.4E-10
COOK 2	216	164	181	177	174	193	191	1.3E-10	2.6E-10
CRYSTAL RIVER 3	216	136	131	179	145	139	191	2.3E-11	4.0E-11
DAVIS-BESSE	191	0	80	186	0	85	196	3.0E-13	4.9E-13
DIABLO CANYON 1	258	186	133	129	199	144	141	4.5E-10	1.0E-09
DIABLO CANYON 2	211	184	196	193	195	207	205	4.3E-10	8.2E-10
FARLEY 1	183	142	180	176	154	197	195	4.9E-11	1.1E-10
FARLEY 2	205	166	210	204	181	230	227	2.3E-10	6.1E-10
FORT CALHOUN	268	199	131	165	213	145	178	9.9E-10	2.2E-09
GINNA	226	0	150	201	0	162	211	4.2E-12	8.0E-12
HADDAM NECK	165	147	153	140	166	173	154	4.6E-11	1.4E-10
INDIAN POINT 2	230	200	212	207	214	226	223	1.1E-09	2.6E-09
INDIAN POINT 3	265	244	244	244	257	257	257	1.6E-08	3.4E-08
KEWAUNEE	277	0	123	239	0	134	255	5.0E-12	1.2E-11
MAINE YANKEE	238	186	191	226	198	203	241	4.7E-10	9.7E-10
MCGUIRE 1	219	128	130	130	136	139	138	1.5E-11	2.4E-11
MCGUIRE 2	141	0	100	-27	0	107	-21	2.5E-13	3.6E-13
MILLSTONE 2	177	133	137	137	142	146	147	2.0E-11	3.4E-11
MILLSTONE 3	134	119	119	119	129	129	129	8.2E-12	1.5E-11
NORTH ANNA 1	184	0	160	110	0	169	122	6.4E-12	1.0E-11
NORTH ANNA 2	220	0	176	140	0	188	152	1.5E-11	2.9E-11
OCONEE 1	214	158	84	181	171	91	193	8.1E-11	1.8E-10
OCONEE 2	273	0	75	187	0	80	199	2.9E-13	5.2E-13

PWR Plant Name	RT_{PTS} at EOL from RVID [°F]	RT at 32 EFPY (EOL) [°F]			RT at 48 EFPY (EOLE) [°F]			TWCF Estimated Using Eq. 11-2	
		RT_{MAX-AW}	RT_{MAX-PL}	RT_{MAX-CW}	RT_{MAX-AW}	RT_{MAX-PL}	RT_{MAX-CW}	32 EFPY	48 EFPY
OCONEE 3	236	0	85	180	0	91	192	2.6E-13	4.6E-13
PALISADES	269	212	190	202	229	206	216	2.2E-09	6.0E-09
PALO VERDE 1	123	83	83	83	90	90	90	9.8E-13	1.5E-12
PALO VERDE 2	78	53	53	53	60	60	60	1.6E-13	2.5E-13
PALO VERDE 3	68	43	43	43	50	50	50	9.0E-14	1.4E-13
POINT BEACH 1	274	171	116	226	181	123	240	1.8E-10	3.4E-10
POINT BEACH 2	288	0	114	217	0	123	230	1.8E-12	3.4E-12
PRAIRIE ISLAND 1	163	0	97	123	0	112	138	2.2E-13	4.9E-13
PRAIRIE ISLAND 2	150	0	93	107	0	108	122	1.8E-13	3.8E-13
ROBINSON 2	255	146	152	196	154	160	209	4.4E-11	7.1E-11
SALEM 1	253	218	225	222	231	238	235	3.4E-09	7.0E-09
SALEM 2	227	166	152	151	180	163	161	1.4E-10	3.1E-10
SEABROOK	120	91	91	91	100	100	100	1.6E-12	2.7E-12
SEQUOYAH 1	235	0	203	150	0	218	164	6.6E-11	1.4E-10
SEQUOYAH 2	152	0	113	81	0	123	90	4.8E-13	8.3E-13
SHEARON HARRIS	196	147	163	162	153	172	170	5.0E-11	7.2E-11
SONGS-2	146	147	147	147	162	162	162	4.5E-11	1.1E-10
SONGS-3	125	110	110	110	122	122	122	4.8E-12	1.0E-11
SOUTH TEXAS 1	84	51	57	57	57	65	65	1.5E-13	2.2E-13
SOUTH TEXAS 2	67	26	31	31	31	37	37	3.3E-14	4.6E-14
ST. LUCIE 1	206	165	150	149	175	159	158	1.3E-10	2.4E-10
ST. LUCIE 2	160	115	115	115	120	120	120	6.5E-12	9.0E-12
SUMMER	113	116	116	116	126	126	126	7.1E-12	1.3E-11
SURRY 1	245	176	145	201	192	161	215	2.5E-10	6.4E-10
SURRY 2	200	152	118	189	164	133	203	5.6E-11	1.2E-10
TMI-1	262	211	73	215	226	80	229	2.0E-09	4.9E-09
TURKEY POINT 3	279	0	102	222	0	108	235	1.9E-12	3.7E-12
TURKEY POINT 4	279	0	96	222	0	103	235	1.8E-12	3.7E-12
VOGTLE 1	118	77	77	-49	82	82	-44	6.6E-13	9.1E-13
VOGTLE 2	126	98	98	98	106	106	106	2.4E-12	3.8E-12
WATERFORD 3	76	69	69	69	77	77	77	4.2E-13	6.6E-13
WATTS BAR 1	253	0	175	97	0	185	106	1.4E-11	2.4E-11
WOLF CREEK	104	81	81	81	87	87	87	8.4E-13	1.2E-12
ZION 1	258	146	102	196	160	115	211	4.1E-11	9.3E-11
ZION 2	272	162	119	225	175	132	241	1.1E-10	2.3E-10

Notes: Plants having a RT_{MAX-AW} value of 0 are forged vessels.
TWCF estimated using Eq. 11-2.
RT_{MAX-AW}, RT_{MAX-PL}, and RT_{MAX-CW} are defined in Eq. 8-1, 8-2, and 8-3, respectively.
RT_{PTS} values taken from [RVID2].

Appendix E – Detailed Reply to Peer Review Comment #22

On the Relevance of Multiple Fracture Initiation-Arrest-Reinitiation Events to the Behavior of Nuclear Reactor Pressure Vessels

Dr. B. Richard Bass
Oak Ridge National Laboratory
Oak Ridge, TN 37831 USA

Dr. Claud E. Pugh
ORSA, Inc.
Knoxville, TN 37922 USA

June 14, 2004

Helmut Schulz's Comment #22:

For the fracture mechanics approach being used the status of validation/verification is well demonstrated for crack initiation and limited stable or unstable crack extension. Although present safety standards or codes do allow the application of crack arrest for multiple initiating events in principle, the supporting experiments are very limited. Furthermore, to the knowledge of the reviewer analysis being performed on such tests (for example NKS test at the MPA Stuttgart) were not able to predict consistently re-initiation and multiple arrest conditions. To evaluate the need to address this issue more deeply, it would be helpful to know if multiple initiation and arrest conditions are really connected to the scenarios being investigated or are only treated as theoretical possibility.

Basic Reply by Staff of Oak Ridge National Laboratory (ORNL)

Given that a flaw may exist within the wall of nuclear reactor pressure vessel (RPV), it is consistent with U.S. experimental evidence and analytical fracture-mechanics predictions that the flaw can propagate into the RPV wall by multiple cleavage run-arrest events when the RPV is exposed to hypothetical pressurized thermal-shock (PTS) loads. Information supporting this view is given in the following pages.

Supporting Information from ORNL Staff

1. Background

Since the late 1960s, the Heavy-Section Steel Technology (HSST) program at ORNL has conducted integrated experimental and analytical studies of the behavior of RPVs in support of the U.S. nuclear safety endeavors. [In 1974, the U.S. Nuclear Regulatory Commission (NRC) was created and charged with regulatory authority to ensure safety within the U.S. nuclear enterprise. The HSST work has been part of the NRC's safety research studies since that time.] The primary purposes of the HSST program have been to: (1) develop a detailed understanding of the fracture behavior of thick-sections of RPV steels over range of conditions prototypical of RPV applications, (2) establish fracture methods (fracture models and associated computer codes) suitable for representing that behavior, and (3) perform and analyze large-scale fracture experiments to validate the applicability of the methods to prototypical RPV conditions. The studies have examined, in technically progressive phases, fracture initiation (brittle and ductile), arrest, re-initiation, mode interaction, environmental effects (e.g., radiation effects), and many associated features (e.g., effects of limited constraint, flaw shape and size, specimen size, multiaxial loads, cladding, weldments, and time-temperature histories). The experiments have employed RPV plate, forging, weldments, and cladding materials.

The two sets of HSST large-scale experiments that have most clearly involved multiple fracture initiation-arrest events have been the thermal-shock experiments (TSEs) of thick cylinders and the nonisothermal wide-plate experiments (WPEs). For purposes of the present discussion, it is felt that the TSEs provide the best vehicle for demonstrating (1) the feasibility of multiple run-arrest fracture events occurring in thick cylindrical sections exposed to thermal transients and (2) the ability of the fracture mechanics models used in the NRC's PTS reevaluation project to predict such behavior. Specifically, applicability of the fracture methodology embodied in the computer program FAVOR (v.03.1) (Dickson 01a, 01b) is the focus here.

The reader is referred to (Pugh 88) and (Bass 86) for information concerning the WPE test series. While the sixteen WPEs experienced multiple run-arrest events, the dynamic nature of those tests does not lend them to providing the easy and obvious argument needed here. The conditions of the WPEs were such that dynamic analyses were needed to adequately model them. The large mass and stiffness of the TSE specimens and also of RPVs eliminates dynamic effects from fracture considerations in those cases.

2. Thermal-Shock Experiments (TSEs)

2.1 Overview of TSE Program

During the 1970s and early 1980s, the HSST program performed a series of 12 fracture experiments using large-scale (1-m diameter) pressure vessels under steady pressure and temperature conditions. Those experiments were called Intermediate Vessel Tests (ITVs), and they are individually discussed in ORNL reports (Bryan 75, 78a, 78b, 79, 85, 87a, 87b). When the ITV test series was well underway, the HSST program planned and started a phase of large-scale fracture experiments using thick cylinders and thermal-shock loads. This series of eight thermal-shock experiments (TSEs) was carried out at ORNL from 1975 to 1983 under the leadership of R. D. Cheverton.

(Cheverton 86) gives an extended summary of these experiments and describes that the TSEs were conducted to investigate the behavior of surface cracks in thick cylinders under conditions relevant to those that could be encountered during a large-break LOCA. It was known that injection of cold water by the emergency core cooling system into a hot reactor vessel after a LOCA would produce low temperature and high thermal-stress conditions under which a small flaw might extend. It was also known that the propensity for crack propagation in an operating RPV would depend upon the degree of fracture-toughness degradation due to neutron exposure and the temperature of the cooling water.

When planning the TSE tests, the ORNL team recognized that thermal-shock situations involve features that had not been adequately examined experimentally at that time, e.g., biaxial stresses, steep gradients in stress and toughness through the wall, variations in these parameters with time, crack arrest in a rising K_I field, reinitiation of crack propagation after arrest, and warm prestressing (WPS). While the combined effect of high K_I and low fracture toughness in the inner-surface region could result in flaw propagation, it was also recognized that the positive gradient in the fracture toughness through the vessel wall provides a mechanism for crack arrest. Thus, a shallow flaw was envisaged to hypothetically initiate, propagate through a distance, arrest, and then reinitiate as the transient time progresses. Consequently, it was expected that deep flaws could result from this type of initiation-arrest scenario and potentially experience multiple initiation and arrest events during a single thermal transient.

As described in (Cheverton 86), ORNL constructed critical-crack-depth curves like that shown in Fig. 1 as an aid in designing and interpreting the TSEs. Such pretest curves were constructed using the OCA-II (see Ball 84) computer code, and posttest analyses used the OCA-P computer code (see Cheverton 84). **These codes were forerunners of the current FAVOR computer code that is being used in the NRC's PTS reevaluation program**.

As an illustration, Fig. 1 shows the predicted behavior of a surface-breaking flaw during a hypothetical transient by plotting the crack depths corresponding to initiation and arrest events ($K_I = K_{Ic}$ and $K_I = K_{Ia}$). Multiple crack run-arrest events are shown in this example computation.

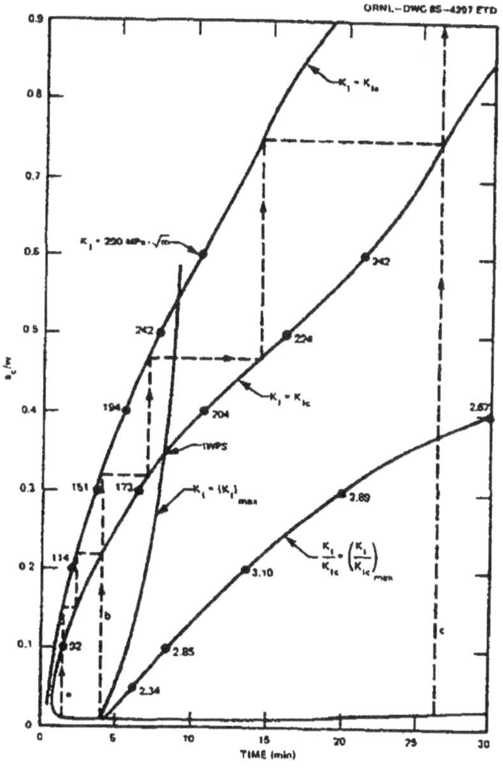

Fig. 1. Critical-crack-depth curves for a PWR vessel during a hypothetical LBLOCA. (a = crack depth and w = wall thickness)

Warm prestressing (WPS) was also recognized to be capable of preventing reinitiation at depths less than the final arrest depth indicated by Fig. 1. The WPS concept means that a flaw will not initiate when K_I is decreasing with time even though K_I may reach or exceed K_{Ic}. Under thermal-shock loading, a deep flaw can conceivably experience its maximum K_I value at a time before the crack-tip temperature has decreased enough to make $K_I \geq K_{Ic}$. During the thermal shock, K_I for a given crack depth increases and then decreases with the temperature gradient. However, K_{Ic} continues to decrease as long as the temperature continues to decrease. The curve in Fig. 1 labeled $K_I = (K_I)_{max}$ shows the times at which K_I reaches a maximum for various crack depths. Thus, for times less than indicated by this curve K_I is always increasing, and for greater times K_I is always decreasing. If WPS is effective, then crack initiation would be limited to times to the left of this curve.

Eight TSE experiments were carried out in two phases that used different specimens and test conditions. The first four TSEs were conducted from September 1975 to January 1977 and used hollow cylindrical specimens fabricated from the trepanned cores taken from the ITV forgings (A508 class-2 steel). These tests are discussed in detail by (Cheverton 76 and 77). The test system used chilled water or water-alcohol mixtures (-23°C) to produce thermal stresses in the heated (288°C) test specimens containing a long internal surface flaw. The test cylinders had an OD of 530 mm (21-in.) and an ID of 240 mm (9.5-in.), and they were 910mm (36-in.) in length. The flaws were shallow with a depth of either 11mm (0.42-in.) or 19 mm (.75-in.). As discussed by (Cheverton 76 and 77), the fracture results from these four experiments were in good agreement with predictions from LEFM analyses which made use of properties values from small laboratory specimens. However, because of the specimen stiffness, deep crack penetrations could not be achieved. Therefore, from August 1979 to May 1983, a second set of four experiments (TSE-5, 5A, 6, and 7) was performed with larger specimens in which deeper crack advances could occur. This second set of four TSEs is addressed in detail in the following pages.

Identification of the Four TSEs that used Large (1-m diam) Specimens

Detailed reports covering the second set of four HSST TSE tests are given by (Cheverton 85a and b), and an overall summary is given in (Cheverton 86). The cylinders used for these tests were made of A508, Class 2 steel, and had an OD of 991mm (39-in.), ID of 682 mm (27-in.), and length of 1220 mm (48-in.). They were given a heat treatment to result in the desired fracture-toughness values. Figure 2a shows a schematic of a TSE test cylinder installed in the ORNL TSE test facility. The test cylinders contained inner-surface flaws and were heated to 96°C prior to being submerged in liquid nitrogen (-196°C) to provide the thermal shock to the inner surface. A specimen and tank schematic are shown in Fig. 2b.

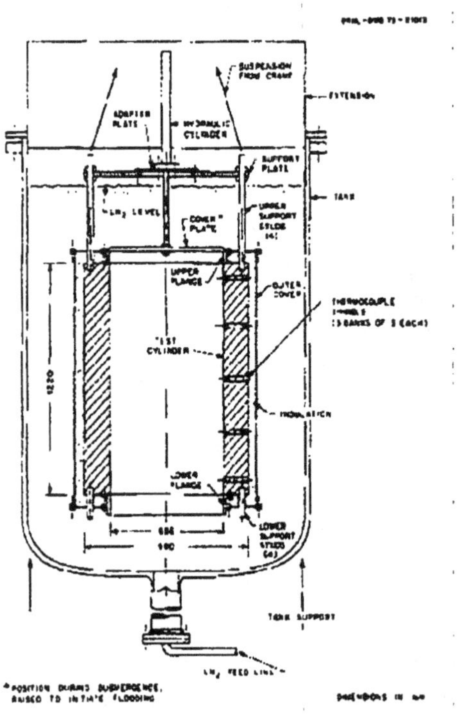

Thermal-shock test cylinder and facility (ORNL, USA).
(ORNL-DWG 78-E1013)

Fig. 2(a). TSE test facility used in TSE-5, 5A, 6, and 7.

Fig. 2(b). Typical TSE test specimen like those used in TSE-5, 5A, 6, and 7.

The test cylinders for TSE-5, 5A, and 6 contained long surface flaws with depths a = 16, 11, and 7.6 mm (0.3-in.), respectively, while the TSE-7 specimen contained a finite length flaw that was 37 mm (1.46-in.) long and 14 mm (0.55-in.) deep. Since the emphasis in this current writing is on multiple crack initiation-arrest-reinitiation events, the following paragraphs address TSE-5, 5A, and 6 where the long flaws propagated into the wall is multiple steps. In TSE-7, the initial flaw extended on the surface in a single event and bifurcated many times to produce an extensive cracking pattern, and therefore this complex situation is excluded from the following discussion. Key observations from experiments TSE-5, 5A, and 6 are given below along with the pretest and posttest analysis results.

2.3. Pretest Analyses of TSE-5, 5A, and 6

In order to further describe the test conditions and results of the pretest analyses that were performed at ORNL, Section 4.2 (*Final Pretest Analysis*) of NUREG/CR-4249 (Cheverton 85a) follows below. The pretest analyses predicted that TSE-5, 5A, and 6 would experience three, four, and one crack initiation-arrest events, respectively. Later in Paragraph 2.4 of this present document, a summary comparison is given of these pretest results, those of posttest analyses, and experimental observations.

The discussion below from (Cheverton 85a) show the critical-crack-depth curves derived from pretest analyses. These curves predicted that TSE-5 would experience three initiation-arrest events, and the final predicted crack depth of a/w = 0.50 to 0.70, depending upon the effectiveness of WPS effects. In the case of TSE-5A, four run-arrest events were predicted with the final a/w ≈ 0.65 (assuming warm prestressing is effective). One fracture initiation event was predicted for TSE-6 with the final a/w ≈ 0.95.

4.2 Final Pretest Analysis

4.2.1 TSE-5

Test conditions assumed for the final pretest analysis of TSE-5 are listed in Table 4.1, and the results of the analysis are presented in Figs. 4.6 and 4.7. Figure 4.6 shows a set of critical-crack-depth curves that indicates the expected behavior of the long axial flaw during TSE-5. It includes the initiation curve ($K_I = K_{Ic}$), the arrest curve ($K_I = K_{Ia}$), the warm prestress curve [$K_I = (K_I)_{max}$], and the maximum K-ratio curve [$K_I/K_{Ic} = (K_I/K_{Ic})_{max}$]. Also included in Fig. 4.6 are temperatures corresponding to several points along the initiation curve and several values of $(K_I/K_{Ic})_{max}$ along the maximum K-ratio curve.

The dashed lines in Fig. 4.6 represent the possible behavior of the flaw during TSE-5, assuming $(a/w)_{initial} = 0.10$. If warm prestressing were not effective and if the crack front remained reasonably uniform in depth (retained its two-dimensional nature), it is apparent that the flaw would penetrate ~90% of the wall. However, if warm prestressing were effective, the maximum penetration might be limited to ~56%. Deeper penetration could result, even with warm prestressing effective, if the dashed curve happened to hit the $K_I = K_{Ic}$ curve just below the point of incipient warm prestressing, in which case the penetration would be ~70%. The least penetration would occur when the dashed line hit the $K_I = K_{Ic}$

Table 4.1. Test conditions for TSE-5 pretest analysis

Test cylinder	TSC-1
Test-cylinder dimensions, m	
Outside diameter	0.991
Inside diameter	0.686
Wall thickness	0.152
Length	1.22
Test-cylinder material	A 508 class-2 chemistry
Test-cylinder heat treatment	Tempered at 613°C for 4 h
K_{Ic} and K_{Ia} curves used in TSE-5 final design analyses	ASME Sect. XI, Appendix A RTNDT = −34°C
Flaw (initial)	Long axial sharp crack, a = 15 mm
Temperatures, °C	
Wall (initial)	93
Sink	−196
Fluid-film heat transfer coefficient	h_f vs T curve based on liquid-nigrogen development studies

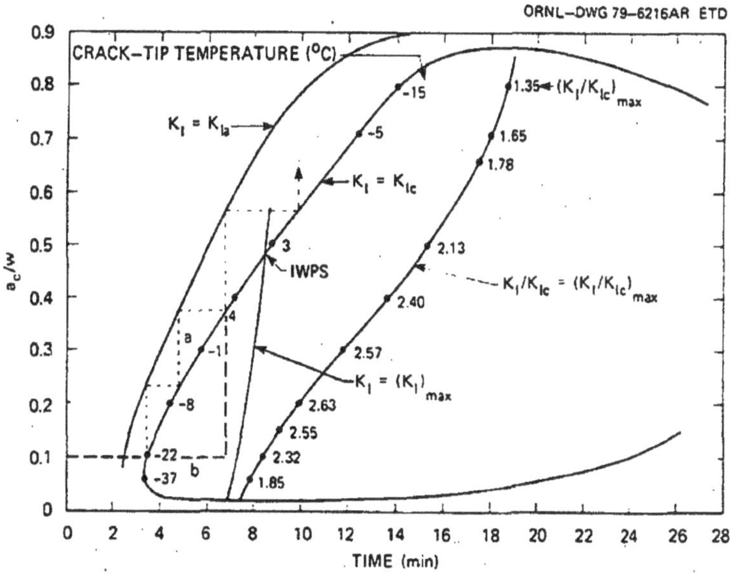

Fig. 4.6. Critical-crack-depth curves for pretest analysis of TSE-5.

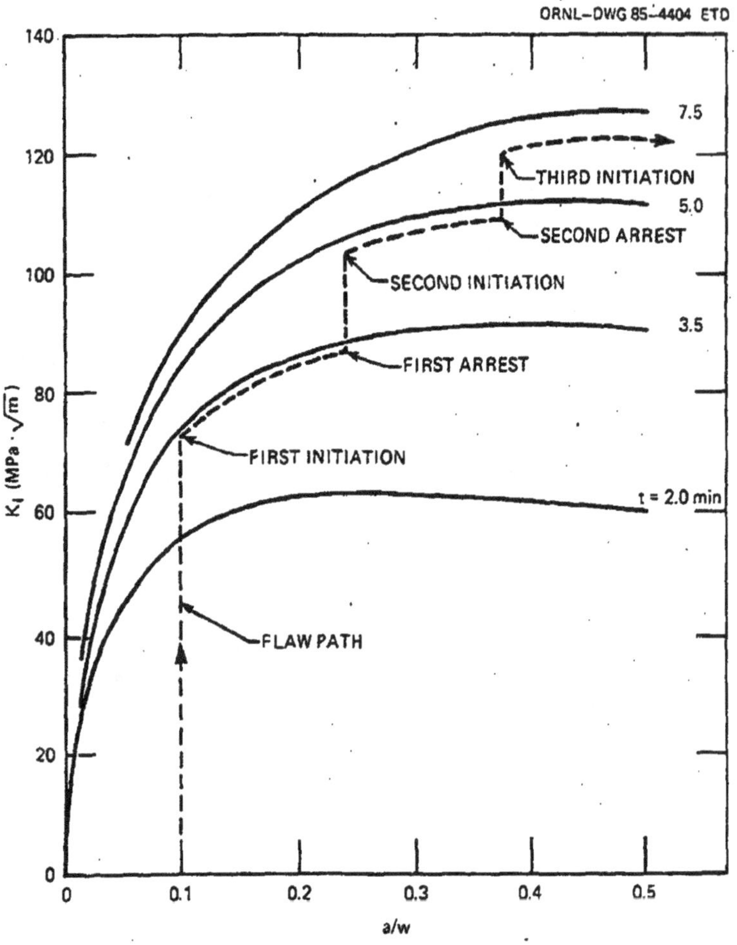

Fig. 4.7. K_I vs a/w for TSE-5 design conditions.

curve just above incipient warm prestressing. For this case the penetration would be limited to ~50%. Thus, the nominal range of predicted final crack depth for TSE-5 was 50 to 70%.

In Fig. 4.6 the area circumscribed by the $K_I = K_{Ic}$ curve and the $K_I = (K_I)_{max}$ curve represents a zone in which flaws with higher-than-anticipated effective toughness could initiate. The result of the higher toughness could be an increase in the crack-jump distance, as illustrated in Fig. 4.6, assuming an initial flaw depth (a/w) of 0.1 and taking the dashed-line path b as opposed to a. In this hypothetical case, crack initiation would be delayed from t = 3.4 min to 6.8 min, and the crack would extend in a single jump from a/w = 0.1 to 0.56 instead of to a/w = 0.23.

It is also of interest to note in Fig. 4.6 that a long crack jump can also be obtained with a very shallow, sharp flaw, that is, with flaw depths represented by the lower portion of the $K_I = K_{Ic}$ curve. Because of the inadvertent inclusion of such a flaw in TSE-5, this is an important observation.

The $(K_I/K_{Ic})_{max}$ curve in Fig. 4.6 represents a measure of the degree of assurance that K_I will actually become equal to K_{Ic} during the experiment. It is important that this condition exist for the experiment to be successful; that is, initiation of flaws with depths less than that corresponding to incipient warm prestressing must take place. Furthermore, to adequately demonstrate warm prestressing, there must be no doubt that the actual value of $(K_I/K_{Ic})_{max}$ for the final flaw depth is greater than unity. Thus, the nominal calculated value of $(K_I/K_{Ic})_{max}$ must be large enough to accommodate all reasonable uncertainties in this ratio and, thus, result in $(K_I/K_{Ic})_{max} > 1$. For the TSE-5 initial flaw, the nominal calculated value of $(K_I/K_{Ic})_{max}$ was ~2.3, and for the deepest final flaw (a/w = 0.7) it was 1.7. These values were considered to be adequate for TSE-5.

Figure 4.7 is a plot of K_I vs a/w for several values of time in the transient. This type of plot is used to determine whether arrest will take place in a rising K_I field. The predicted path (K_I vs a/w and t) for the first three initiation-arrest events is shown with the dashed lines and is based on data taken from Fig. 4.6. As indicated, the first two initiation-arrest events take place with positive values of $dK_I/d(a/w)$, while the third arrest event takes place with a negative value of $dK_I/d(a/w)$. Thus, the predicted behavior for the first two arrest events is consistent with the objective of demonstrating arrest in a rising K_I field.

It is observed in Fig. 4.7 that the K_I values corresponding to initiation and arrest events are in the range of 70 to 130 $MPa\cdot\sqrt{m}$. As mentioned in Chap. 1, it was hoped that a maximum critical K_I value near 250 $MPa\cdot\sqrt{m}$ could be achieved. With this exception, the results of the final pretest analysis indicated that all of the objectives for TSE-5 could be met.

4.2.2 TSE-5A

Because the scope of TSE-5A was the same as that for TSE-5, the pretest calculated behavior of the long axial flaw for the two experiments was similar. However, there were differences in the assumed thermal transients, toughness curves, and initial flaw depths, and these differences resulted in somewhat different calculated behaviors for the flaws.

Test conditions assumed and/or specified for TSE-5A are shown in Table 4.2 and include a thermal transient equal to that actually achieved during TSE-5 (more severe than used in the TSE-5 pretest analysis), toughness curves based on very extensive material-characterization studies (lower-bound K_J and mean K_{Ia}), and an initial crack depth (a/w) equal to 0.075 as compared with 0.10 for TSE-5.

Results of the final pretest analysis for TSE-5A are shown in Fig. 4.8. As indicated by this set of critical-crack-depth curves, the flaw would penetrate deeper than 90% of the wall without warm prestressing but only 45 to 65% with warm prestressing. The range of $(K_I/K_{Ic})_{max}$ for the latter range of final crack depths is 2.2 to 1.7, and, thus, presumably

Table 4.2. Test conditions for TSE-5A pretest analysis

Test cylinder	TSC-2
Test-cylinder dimensions, m	
Outside diameter	0.991
Inside diameter	0.686
Wall thickness	0.152
Length	1.22
Test-cylinder material	A 508 class-2 chemistry
Test-cylinder heat treatment	Tempered at 679°C for 4 h
K_{Ic} and K_{Ia} curves used in design	Lower-bound K_J and mean K_{Ia} data from ORNL and BCL TSE-5A material-characterization studies (see Fig. 4.4)
Flaw (initial)	Long axial sharp crack, a = 11 mm
Temperatures, °C	
Wall (initial)	93
Sink	−196
Fluid-film heat transfer coefficient	h_f vs T curve from TSE-5

an adequate demonstration of warm prestressing could be achieved. Assuming warm prestressing to be effective, there would be four initiation-arrest events, and the longest crack jump ($\Delta a/w$) would be the last and equal to 0.20.

Figure 4.8 includes a curve referred to as $K_I = (K_I)_{max(a/w)}$, which represents the crack depth for which $dK_I/da = 0$ for different times in the transient. If arrest takes place at a depth less than that corresponding to the $K_I = (K_I)_{max(a/w)}$ curve, it will do so in a rising K_I field. Thus, according to the pretest analysis, the first two arrest events would take place in a rising K_I field.

Another point of interest illustrated in Fig 4.8 is that the K_I values corresponding to initiation and arrest events cover a range from ~70 to 140 $MPa\cdot\sqrt{m}$, that is, the lower to midtransition region. This range is similar to that calculated for some PWR large-break loss-of-coolant accident cases, but the maximum value is still less than desired (250 $MPa\cdot\sqrt{m}$). With this exception, the results of the final pretest analysis indicated that the objectives of TSE-5A could be met.

4.2.3 TSE-6

A long crack jump with arrest deep in the wall was specified for TSE-6. To achieve this without the initial flaw having to be very shallow or blunted, it was necessary to use a thinner wall (76 mm) for TSE-6

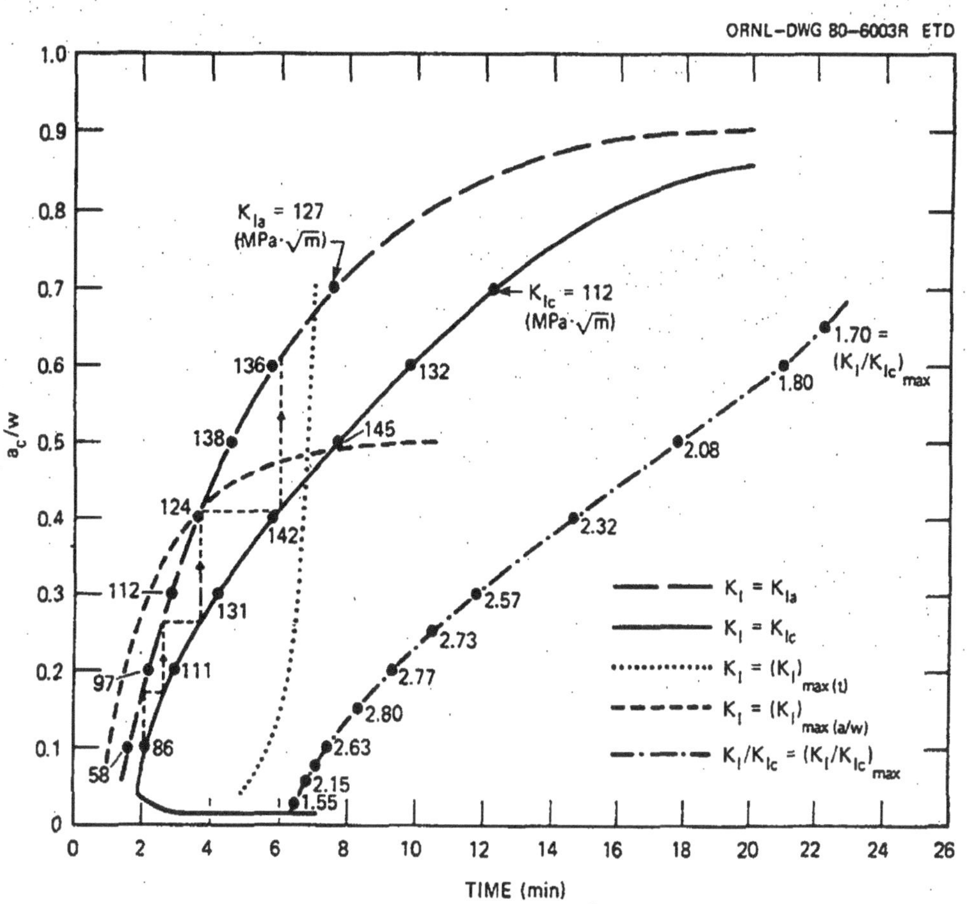

Fig. 4.8. Pretest critical-crack-depth curves for TSE-5A assuming TSE-5 thermal transient and TSE-5A lower-bound K_{IJ} and mean K_{Ia} curves.

than that used for the TSE-5 and TSE-5A test cylinders (152 mm) and to use the same low fracture-toughness properties actually obtained for the TSE-5 test cylinder (see Table 4.3 for test conditions used in pretest analysis). This combination resulted in a set of calculated critical-crack-depth curves with nearly vertical sections of the initiation and arrest curves (Fig. 4.9). The dashed lines in Fig. 4.9 represent the expected behavior of the flaw and indicate a single long crack jump from a/w = 0.1 (initial flaw) to a/w = 0.95. Reinitiation would not be likely because of warm prestressing or $(K_I/K_{Ic})_{max} < 1$. For the initial flaw, $(K_I/K_{Ic})_{wps} = 1.36$, which was large enough to ensure initiation prior to warm prestressing. Furthermore, K_I values all along the initiation and arrest curves were below upper-shelf toughness values.

The expected inability of the flaw to completely penetrate the wall under thermal-shock loading conditions only is illustrated in Fig. 4.10,

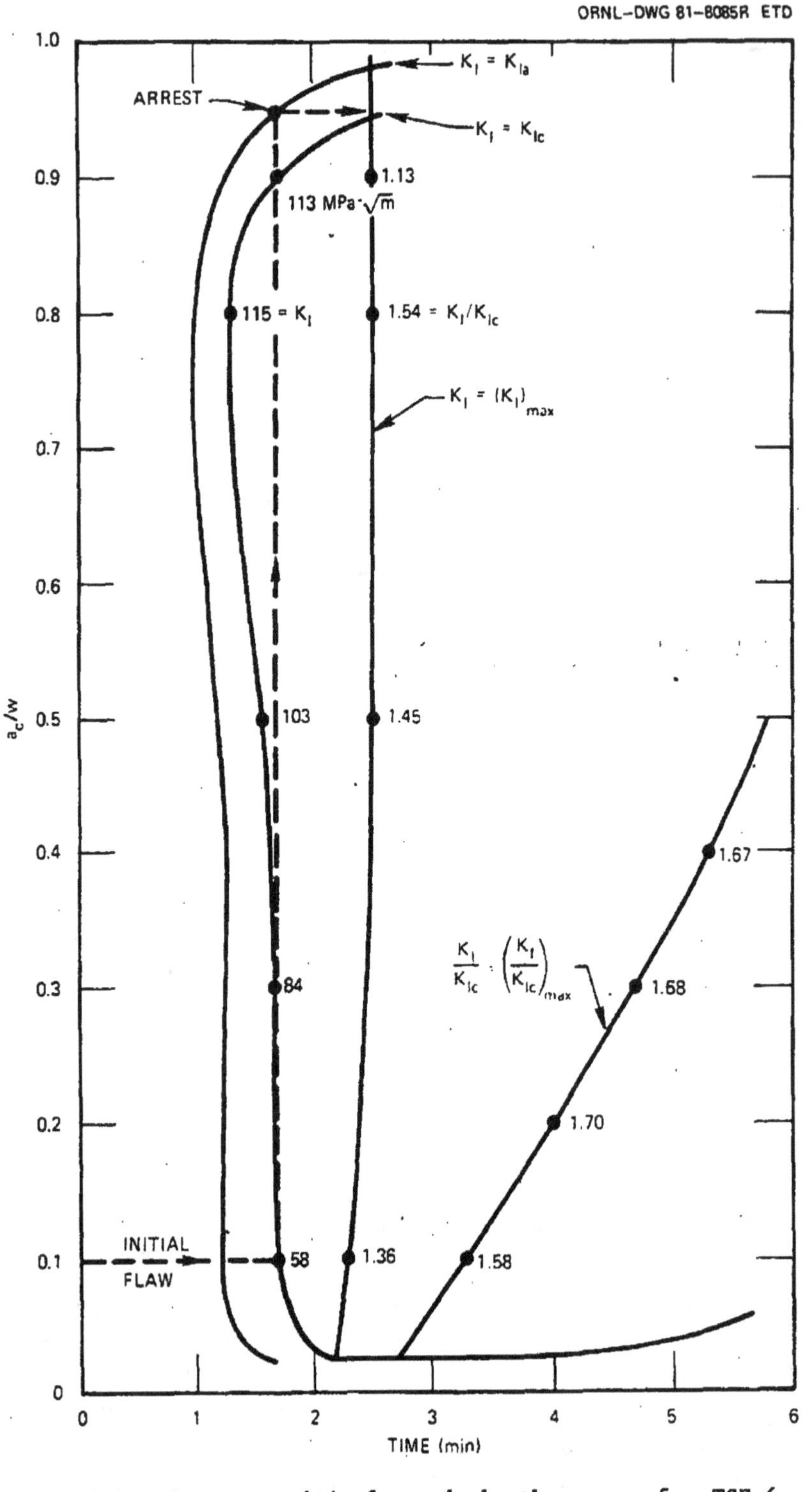

Fig. 4.9. Pretest critical-crack-depth curves for TSE-6.

Table 4.3. Test conditions for TSE-6 pretest analysis

Test cylinder	TSC-3
Test-cylinder dimensions, m	
Outside diameter	0.991
Inside diameter	0.838
Wall thickness	0.076
Length	1.22
Test-cylinder material	A 508 class-2 chemistry
Test-cylinder heat treatment	Tempered at 613°C for 4 h
K_{Ic} and K_{Ia} curves used in design	Toughness curves deduced from TSE-5 (see Fig. 4.5)
Flaw (initial)	Long axial sharp crack, a = 7.6 mm
Temperatures, °C	
Wall (initial)	93
Sink	−196
Fluid-film heat transfer coefficient	h_f vs T_s deduced from TSE-5A

which is a plot of K_I and K_{Ia} vs a/w at approximately the predicted time of crack propagation (t ≈ 1.7 min). This figure shows the tendency for a steep negative gradient in K_I near the outside surface, with K_I dropping below K_{Ia}. This tendency for linear-elastic fracture mechanics to predict that K_I will approach zero is the result of (1) a limited extent of crack-surface rotation associated with the net thermal bending moment, (2) the resistance of the cylinder wall to bending, and (3) a compressive load on the free-body remaining ligament (see Fig. 4.11). As the crack tip approaches the back surface, the crack-surface rotation and the compressive load approach their maximum values, while the bending moment at the ligament approaches zero. The result is a substantial compressive-stress component in the ligament, and, thus, presumably the crack could not completely penetrate the wall.

The resistance of the wall to bending decreases with an increase in the ratio of cylinder radius to wall thickness (r/w). As shown in Table 4.4, the ratio is much larger for a typical PWR vessel than it is for the thermal-shock test cylinders. Thus, there is a greater tendency for deep propagation of a flaw in the PWR vessels than in the test cylinders. However, a linear-elastic fracture-mechanics analysis of the PWR vessel indicates that for this case, too, a long axial flaw could not fully penetrate the wall under thermal-shock loading only (see Chap. 1). Thus, if during TSE-6 the flaw did not breach the wall, this would be considered an adequate demonstration.

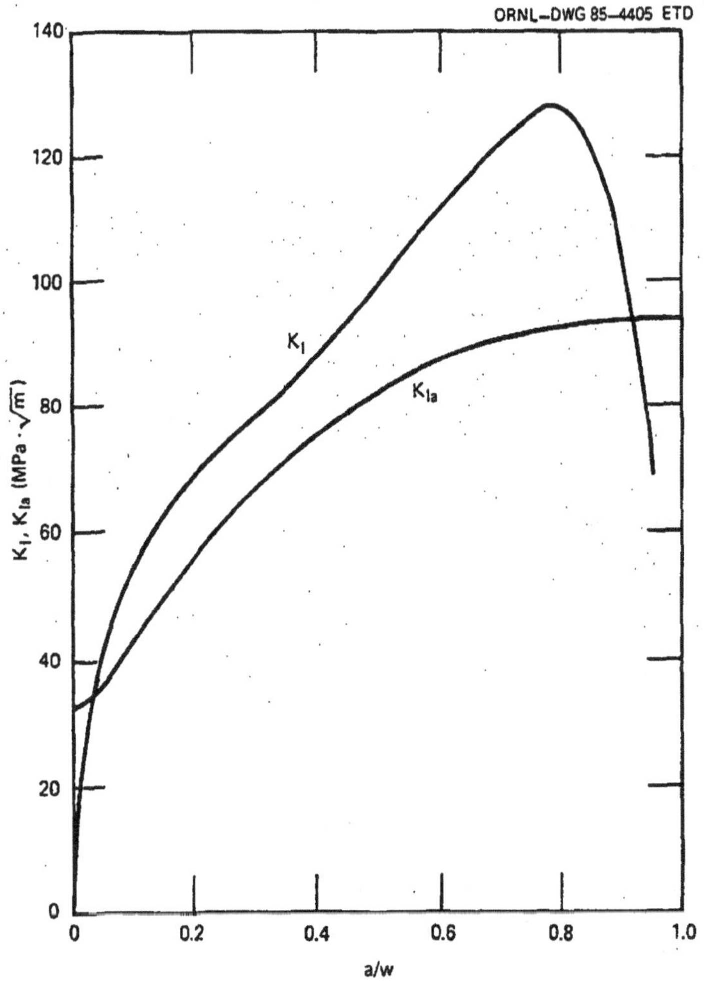

Fig. 4.10. K_I and K_{Ia} vs a/w for TSE-6 at t = 1.5 min (pretest analysis).

Table 4.4. Comparison of radius-to-wall thickness ratios for PWR vessels and ORNL thermal-shock test cylinders

Cylinder	r_i/w^a
PWR vessel	2180/216 = 10
Test cylinders	
TSE-6	419/76 = 5.5
TSE-5, TSE-5A	346/152 = 2.3
TSE-1, TSE-2, TSE-3, TSE-4	114/152 = 0.8

[a]Inside radius (r_i)/wall thickness (w), dimensions in millimeters.

2.4. Observations from Tests TSE-5, 5A, and 6

2.4.1. TSE-5

TSE-5 experienced three initiation/arrest events with deep penetration of the two dimensional flaw. The following pages are taken from Chapter 8 of NUREG/CR-4249 (Cheverton 85a) and address the test results and critical crack-depth curves generated posttest. The posttest analyses used improved fracture-toughness models and actual temperature transients. While the pretest analyses predicted the correct number of initiation-arrest events, the posttest analyses provided improved results.

8.3 Crack Depths

Estimates of crack depth prior to the test and at the time of each of the three events were obtained from three UT transducers secured to the outer surface of the test cylinder directly opposite the tip of the flaw. Once the test was completed and the outer-surface insulation was removed, UT instrumentation was used to obtain a detailed description of the final crack depth. Finally, a pie-shaped section of the cylinder wall containing the primary flaw was removed from the cylinder and was cut into several lengths to expose cross sections of the flaw at several positions along the length of the cylinder (Fig. 8.9). Once the cross sections were examined, the fracture in each block was completed under mechanical loading conditions at cryogenic temperatures to reveal the fracture surfaces. The cross sections and fracture surfaces revealed the initial crack depth, first crack-arrest depth, and final crack depth, but the second arrested crack depth was not discernible. A typical cross section and fracture surface are shown in Figs. 8.10 and 8.11, and all depth measurements are shown in Fig. 8.12.

The data in Fig. 8.12 show good agreement between crack-depth measurements and indicate that crack penetration in the central portion of

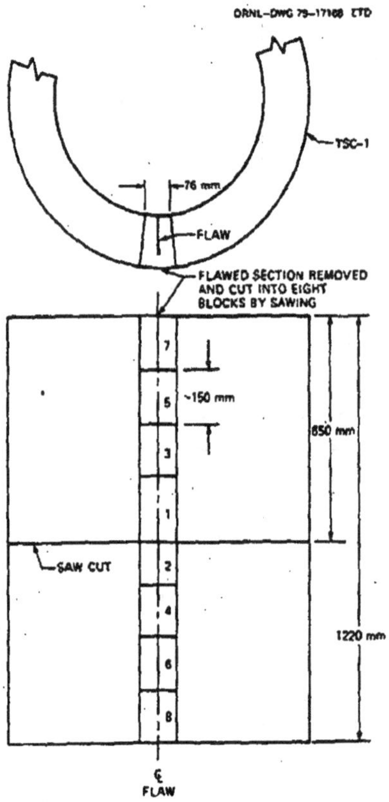

Fig. 8.9. Section of TSE-5 test cylinder removed for study of crack profile and fracture surfaces.

M&C PHOTO Y164626

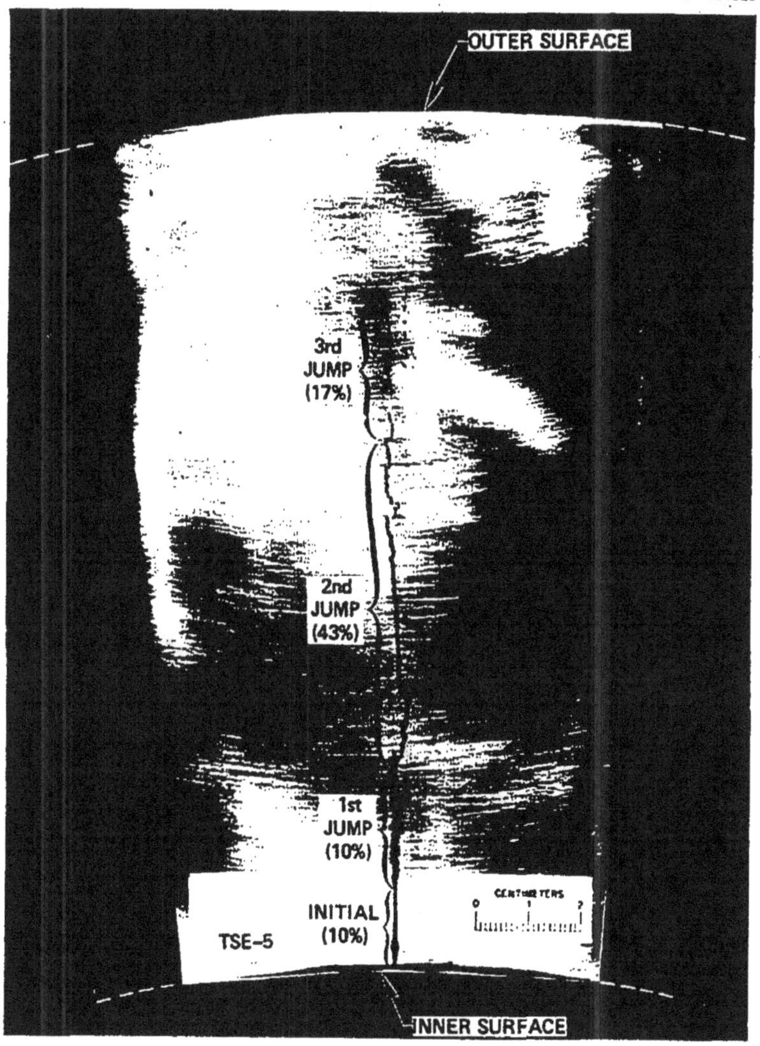

Fig. 8.10. Cross section of TSE-5 final flaw near midlength of test cylinder.

8.4 Fracture-Mechanics Analysis of Primary Flaw

8.4.1 Posttest analysis based on design toughness curve

The TSE-5 pretest fracture-mechanics analysis indicated that three initiation-arrest events would take place during TSE-5, and indeed that is what happened. However, a comparison of the estimated and actual times for the events shows that the latter times were significantly earlier due in part to the greater severity of the actual thermal shock. Thus, before additional comparisons could be made, it was necessary to conduct a posttest fracture-mechanics analysis using the measured temperature distributions. The corresponding critical-crack-depth curves, assuming that the TSE-5-design K_{Ic} vs temperature curve was appropriate, are shown in Fig. 8.13. Superimposed on these curves is the "actual"

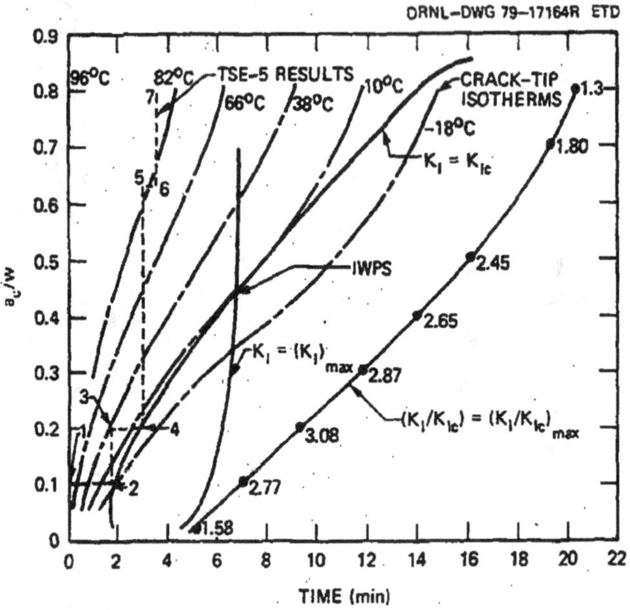

Fig. 8.13. Critical-crack-depth curves from posttest analysis of TSE-5 (original K_{Ic} data).

path of events, using the measured midlength crack-arrest depths. As indicated, the first two initiation events agree well with the calculations, but the very long second crack jump [$\Delta(a/w) \simeq 0.4$] resulted in arrest at a much greater depth and temperature than expected. Thus, the toughness at the higher temperature had to be much less than indicated by the design-toughness curve.

8.4.2 K_{Ic} and K_{Ia} values deduced from TSE-5

The calculated K_I vs time and K_I vs a/w curves for the actual thermal shock are shown in Figs. 8.14 and 8.15 with the initiation-arrest events superimposed. As indicated in these figures, the K_I values corresponding to the three initiation-arrest events are 79, 111, and 115 MPa·$\sqrt{m}$ for the initiation events and 86, 104, and 92 MPa·$\sqrt{m}$ for the arrest events. Corresponding temperatures obtained from Figs. 8.1–8.3 are –9, –3, and 79°C for the initiation events and 36, 82, and 89°C for the arrest events. All of these data are summarized in Table 8.1.

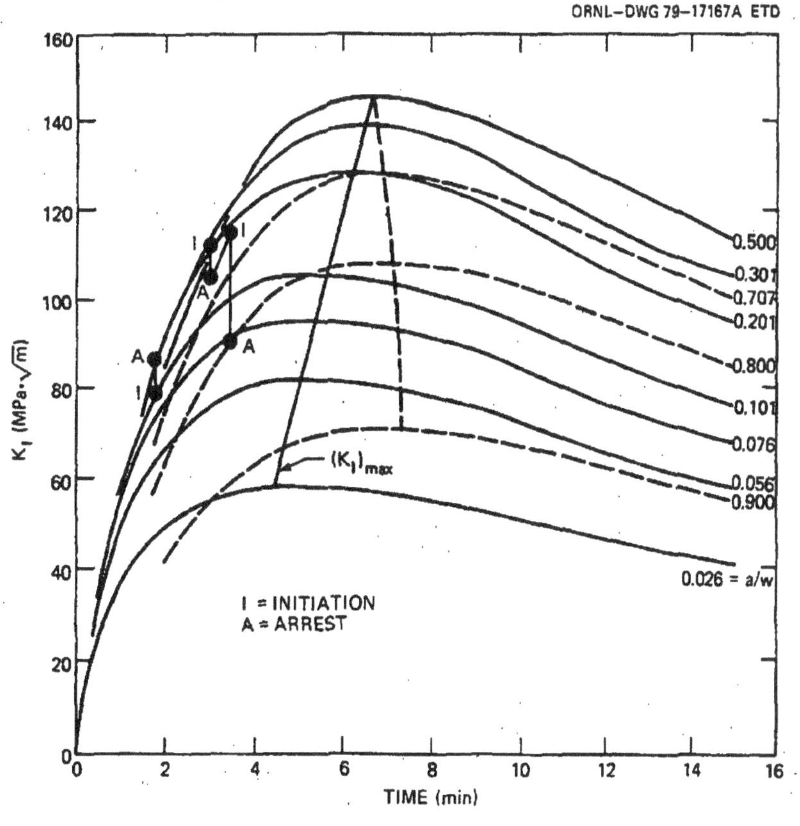

Fig. 8.14. K_I vs time for several crack depths (TSE-5 posttest analysis).

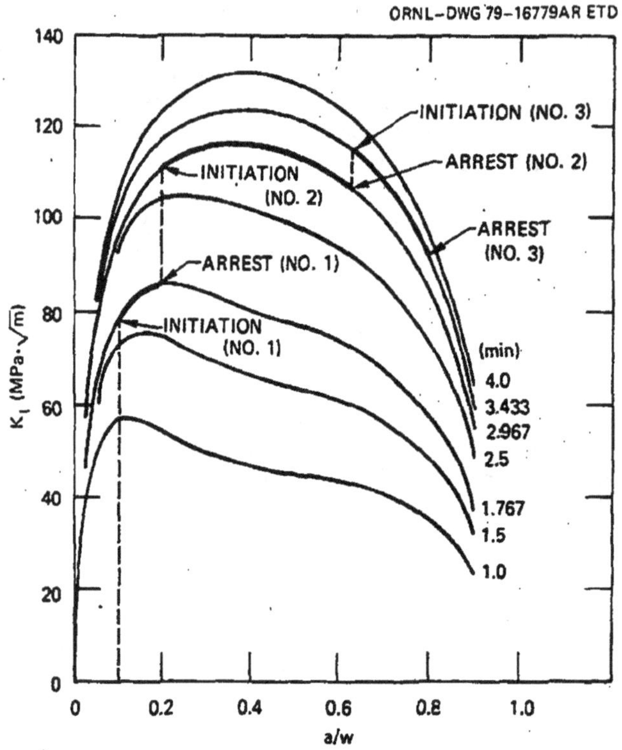

Fig. 8.15. K_I vs a/w for several times in the TSE-5 transient (posttest analysis).

Table 8.1. Summary of data corresponding to the events for the long axial flaw (TSE-5)

Parameters	Initiation-arrest event		
	1	2	3
Time, s	105	177	205
Crack depth,[a] a/w			
Initiation	0.10	0.20	0.63
Arrest	0.20	0.63	0.80
Temperature, °C			
Initiation	−9	−3	79
Arrest	36	82	89
K_{Ic}, MPa·√m	79	111	115
K_{Ia}, MPa·√m	86	104	92
Duration of experiment, min	30		

[a]Maximum depth (midlength of test cylinder).

ORNL–DWG 85–4422 ETD

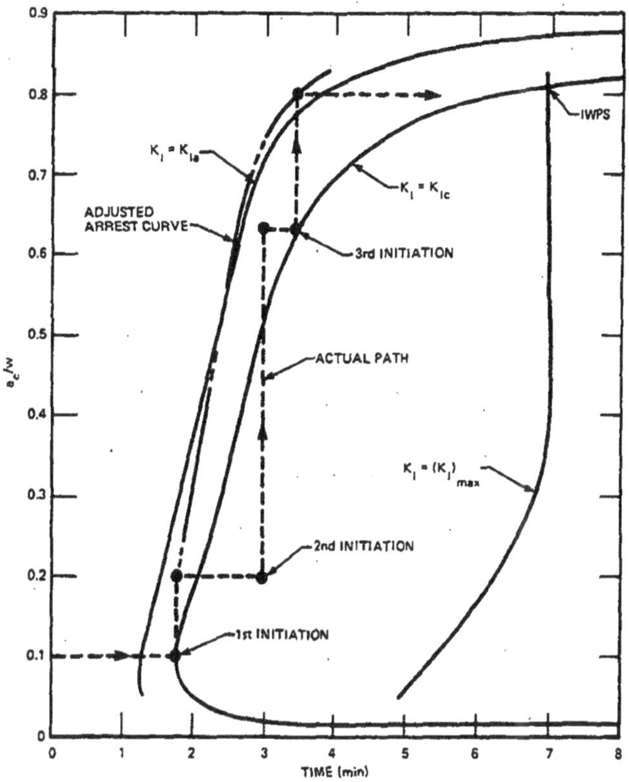

Fig. 8.17. Critical-crack-depth curves from posttest analysis of TSE-5 using TSE-5 K_{Ic} data.

Table 8.2 Crack depths derived from COD data and from direct measurements

Time (s)	Fractional crack depth (a/w)	
	Direct	COD
106	0.10[a]	0.12
106	0.20[a]	0.21
178	0.20[a]	0.22
178	0.63[b]	0.55
206	0.63[b]	0.55
206	0.80[a]	0.72

[a]Fracture-surface measurement.

[b]UT measurement.

In summary, posttest analysis predictions, posttest crack-depth measurements from the fracture surface, and cross-section photo results show good agreement. In conclusion, the LEFM analyses predicted well the nature and magnitude of the multiple-event fracture behavior of TSE-5 under thermal shock loads.

2.4.2. TSE-5A

As discussed on page 34 of (Cheverton 85a), TSE-5A had the same scope as TSE-5, but different thermal transients, fracture toughness, and initial crack-depth made the expected response different for the two experiments. Accordingly, TSE-5A experienced four initiation/arrest events with 50% penetration of the wall, whereas TSE-5 experienced three initiation/arrest events as discussed in the previous paragraphs. A fifth event was prevented by WPS effects, and one of the arrest events took place with K_I increasing with crack depth. After the WPS intervention, the K_I/K_{Ic} ratio reached a maximum value of 2.3 without crack initiation taking place. As was the case for TSE-5, the behavior was consistent in nature with the pretest analyses based on LEFM methods.

The following pages from (Cheverton 85a) discuss the observed crack propagation events and the results of posttest analyses. The posttest analyses here were made using the OCA-P computer code, the best interpretation of actual test specimen material properties and the actual thermal transient. The posttest analyses produce very good predictions of what was observed from test measurements. (Note: When considering Table 9.1 below, it should be kept in mind that the specimen wall thickness was 152 mm. Using this with the crack depth values shown in that table allows calculation of a/w values for ease of comparison with a/w values reported in Table 9.2.)

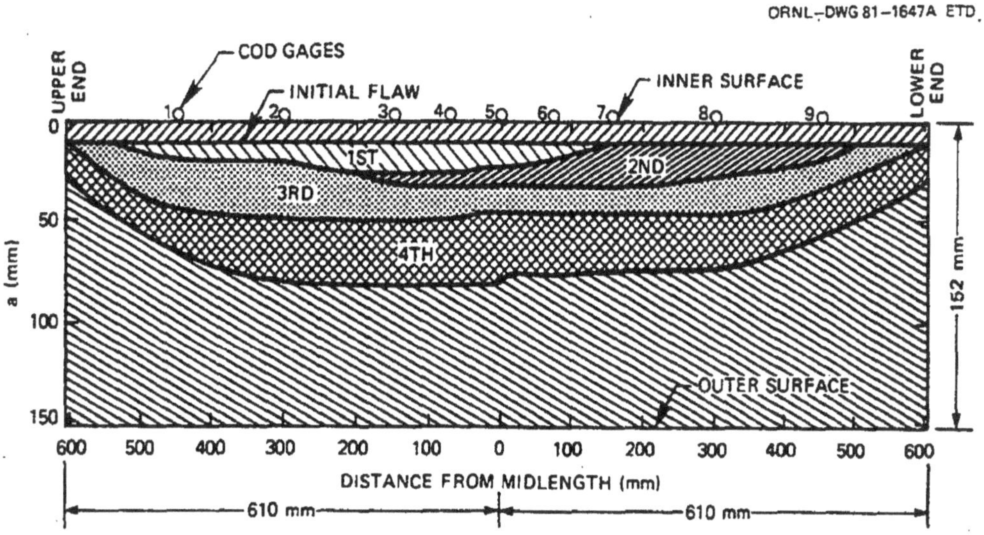

Fig. 9.9. Schematic cross section of TSE-5A test-cylinder wall, showing four-step progression of long axial flaw as deduced from fracture surfaces.

shortly thereafter. A composite photograph of the entire fracture surface is shown in Fig. 9.10 along with an enlarged view of one section. Figure 9.9 was constructed by obtaining measurements of crack depth from photographs of this type.

Figure 9.10 shows a clear indication of an additional event just ahead of the final arrest event. This event was not detected with the COD gages but, rather with the UT instrumentation, as were the four major events. According to an analysis of the UT data, the time between the final arrest event and the preceding arrest event was only 900 µs. This short period of time suggests that the additional event was the result of dynamic effects that created an oscillation in K_I. When K_I dipped down, arrest took place, but immediately thereafter initiation took place as K_I increased. Such oscillations have been observed in lab K_{Ia} tests,[1] and dynamic analyses of the lab specimens indicate such behavior.[2] A dynamic analysis was also attempted for TSE-5A; it did not indicate that a momentary arrest event would take place.[3]

9.3 Crack Depths

Crack depths were determined from the COD and UT data and also from direct observations of the fracture surfaces. The COD data were analyzed in the manner described in Sect. 8.5, and the calculated relation between COD and crack depth for times of interest during TSE-5A is shown in Fig. 9.11. Crack depths based on COD and UT data and obtained from examination of the fracture surfaces are presented in Table 9.1 for the central portion of the test cylinder. As indicated, the agreement is reasonably good.

Table 9.1. Estimated crack depths near midlength of test cylinder for TSE-5A

Event No.	Event	Time (s)	Crack depth (mm)		
			COD	Fracture surface	UT
1	Initiation	78.5	12	12	11
	Arrest		21	23	17
2	Arrest	90.5	30	30	31
3	Arrest	123.0	48	48	41
4	Arrest	184.5	81	81	81

9.4 Fracture-Mechanics Analysis

9.4.1 Posttest analysis based on design toughness curves

The number of initiation-arrest events taking place during TSE-5A agrees with the pretest analysis (Fig. 4.8), but the times at which the events actually took place were earlier than calculated. This indicated that the thermal shock was more severe than that used for the pretest analysis and/or that the actual material toughness was less than that used in the pretest analysis. As planned (Fig. 9.1), the thermal shock was more severe than that assumed for the pretest analysis (that achieved during TSE-5), and, apparently, the toughness was less than assumed, as discussed below.

The toughness curves used in the pretest analysis are shown in Fig. 9.13. The K_{Ic} curve represents the lower bound of fifty 1T-CS data points, while the K_{Ia} curve is a mean curve through six data points obtained from 25 × 151 × 151-mm, wedge-loaded, crack-arrest specimens (also refer to Fig. 7.11).

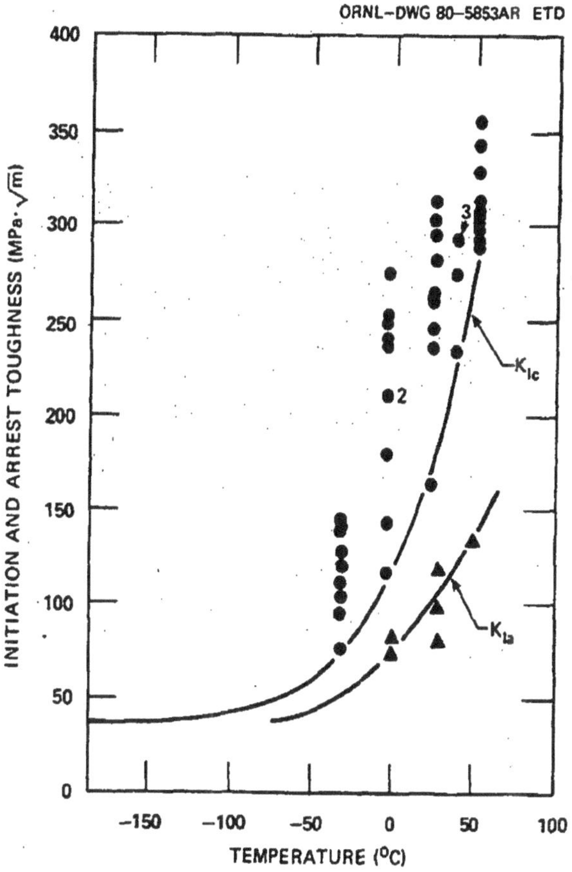

Fig. 9.13. Lower-bound K_J and mean K_{Ia} curves (based on pretest lab data only) used for K_{Ic} and K_{Ia} in TSE-5A pretest analysis.

Posttest analyses of TSE-5A were conducted using the measured test-cylinder temperatures shown in Figs. 9.2–9.6. The results of such an analysis, based on the toughness curves in Fig. 9.13, are shown in Fig. 9.14, which is a set of critical-crack-depth curves that includes the actual path of events. It is observed that the initiation and arrest events fall to the left of their respective initiation and arrest curves, indicating that the actual crack-initiation and crack-arrest toughness values were below the curves in Fig. 9.13.

9.4.2 K_{Ic} and K_{Ia} values deduced from TSE-5A

Critical values of K_I corresponding to the four initiation and arrest events during TSE-5A are shown in Table 9.2, and these values are compared with the experiment-design toughness curves in Fig. 9.15. It is obvious that the K_{Ic} and K_{Ia} values deduced from TSE-5A are substantially less than the values assumed for the design of the experiment on the bases of the lab data. Of course, this was to be expected insofar as crack initiation is concerned because none of the lab initiation data

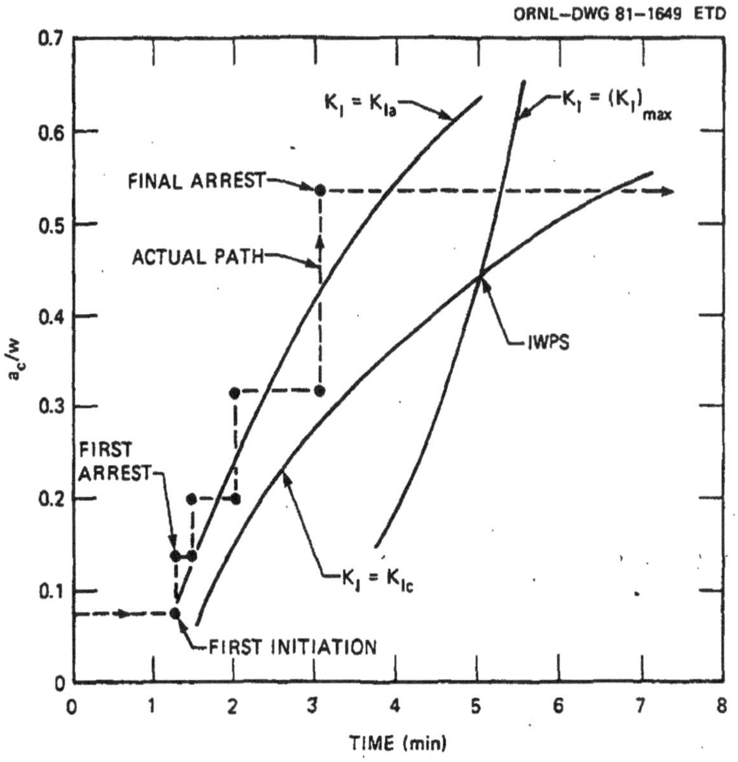

Fig. 9.14. Critical-crack-depth curves for TSE-5A based on toughness curves in Fig. 9.13.

Table 9.2. Summary of critical data for TSE-5A

Time (s)	Event	a/w	K_I (MPa•$\sqrt{m}$)	Temperature (°C)
78.5	1st initiation	0.076	70	−11
90.5	2nd initiation	0.138	85	12
123.0	3rd initiation	0.198	108	13
184.5	4th initiation	0.316	135	21
78.5	1st arrest	0.138	76	22
90.5	2nd arrest	0.198	86	38
123.0	3rd arrest	0.316	107	51
184.5[a]	4th arrest	0.535	130	67

[a] K_I/K_{Ic} reached a maximum value of 2.3 at ~14 min.

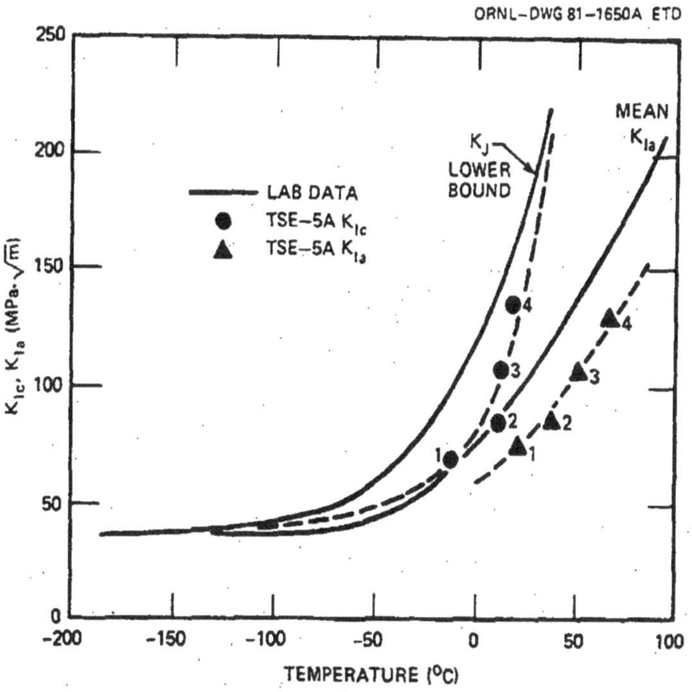

Fig. 9.15. Comparison of K_{Ic} and K_{Ia} values deduced from TSE-5A with laboratory data presented in Fig. 9.13.

points was valid. However, as indicated below, the lab data were adequate for designing the experiment; that is, all objectives of TSE-5A were achieved.

9.4.3 Posttest analysis based on K_{Ic} and K_{Ia} curves deduced from TSE-5A

The dashed curves in Fig. 9.15 represent a best fit of the K_{Ic} and K_{Ia} data points deduced from TSE-5A. These curves were used in a second posttest analysis, and as one would expect, the agreement between actual and "predicted" flaw behavior is much better than when using the solid curves in Fig. 9.15. Results of the second posttest analysis are shown in Fig. 9.16, and a complete set of digital output is included in Appendix E.

9.4.4 Warm prestressing

Figure 9.16 indicates that a fifth crack-initiation event was prevented by warm prestressing, and the maximum value of K_I/K_{Ic} for the final crack depth was 2.3. This maximum value was reached ~9 min after the time of incipient warm prestressing, and the corresponding crack-tip

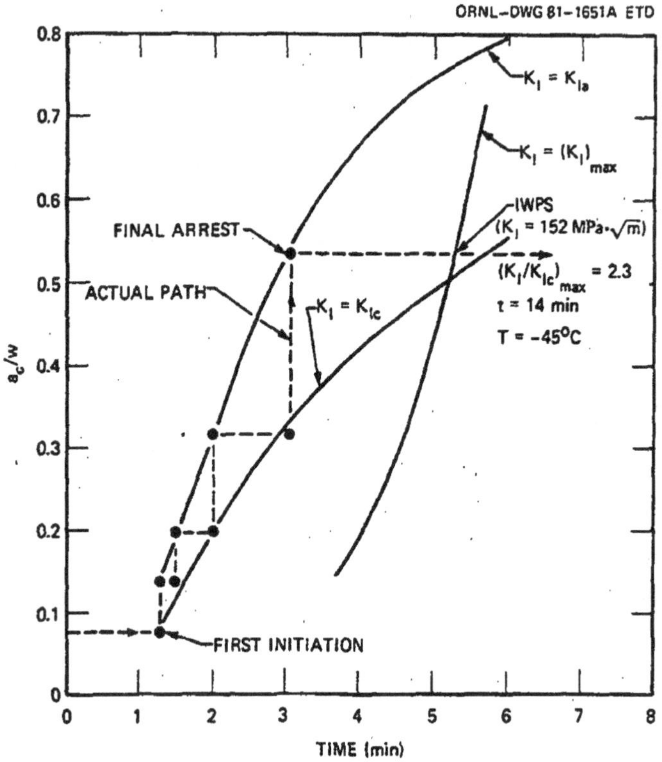

Fig. 9.16. Posttest critical-crack-depth curves for TSE-5A using modified toughness curves in Fig. 9.15.

temperature was −45°C (RTNDT = 10°C). The maximum value of K_I for the final crack depth was 152 MPa·$\sqrt{m}$, and, of course, it occurred at the time of incipient warm prestressing (~5 min).

The value of 2.3 for $(K_I/K_{Ic})_{max}$ is large enough, relative to unity, to compensate for all uncertainties in the experiment and related post-test analysis, and the crack-tip temperature corresponding to this value was well below RTNDT. Thus, even though the value of $(K_I)_{max}$ for the final crack depth was substantially less than that predicted for the PWR LBLOCA, TSE-5A provided a convincing demonstration of the ability of warm prestressing to prevent crack initiation.

2.4.3. TSE-6

The TSE-6 cylinder had a thinner wall (76 mm vs. 152 mm for the other tests) and introduced the potential for a single long crack jump to a depth greater than 90% of the wall thickness. There were actually two crack jumps in the test. The first was relatively short, and the total penetration was 93% of the wall thickness. One contributor to the difference between pretest analyses and test results is that the specimen toughness was somewhat lower than initially assumed. The test results and posttest analyses are discussed in the following pages taken from (Cheverton 85a). In addition to helping demonstrate the predictability of fracture behavior under thermal shock conditions, TSE-6 helped demonstrate the inability of a long flaw to fully penetrate the vessel wall under thermal shock only loads.

10.2 Events During TSE-6

As indicated in Fig. 10.4, which shows the final profile of the flaw near midlength of the test cylinder, the TSE-6 flaw penetrated very deep (>90%) into the wall of the test cylinder but did not extend all the way through, consistent with predictions. The COD data and a photograph of the full length of the fracture surface (Fig. 10.5) indicate that there were three initiation-arrest events involved. Two of these events, the first and last, were detected with COD gages that were connected to "slow" recorders, and the recorded output of the gages is shown in Fig. 10.6. The second event was detected with two COD gages that were connected to a fast-phenomena recorder, and the output for one of these gages is shown in Fig. 10.7. It is apparent that the second arrest event was the result of a momentary decrease in K_I because ~300 μs thereafter the third initiation-arrest event took place, presumably the result of a momentary increase in K_I (the time between arrest events was ~900 μs, the same as for TSE-5A). A dynamic analysis similar to that performed for TSE-5A was also attempted for TSE-6 (Sect. 9.2). This time the analysis indicated that the momentary arrest event would take place.[1]

The TSE-6 dynamic event appears to be the same type of dynamic event that took place during TSE-5A, and it is of interest to note that in TSE-5A the crack jump preceding the short duration of the arrest event was much shorter and the vessel wall was much stiffer. Thus, it is not clear

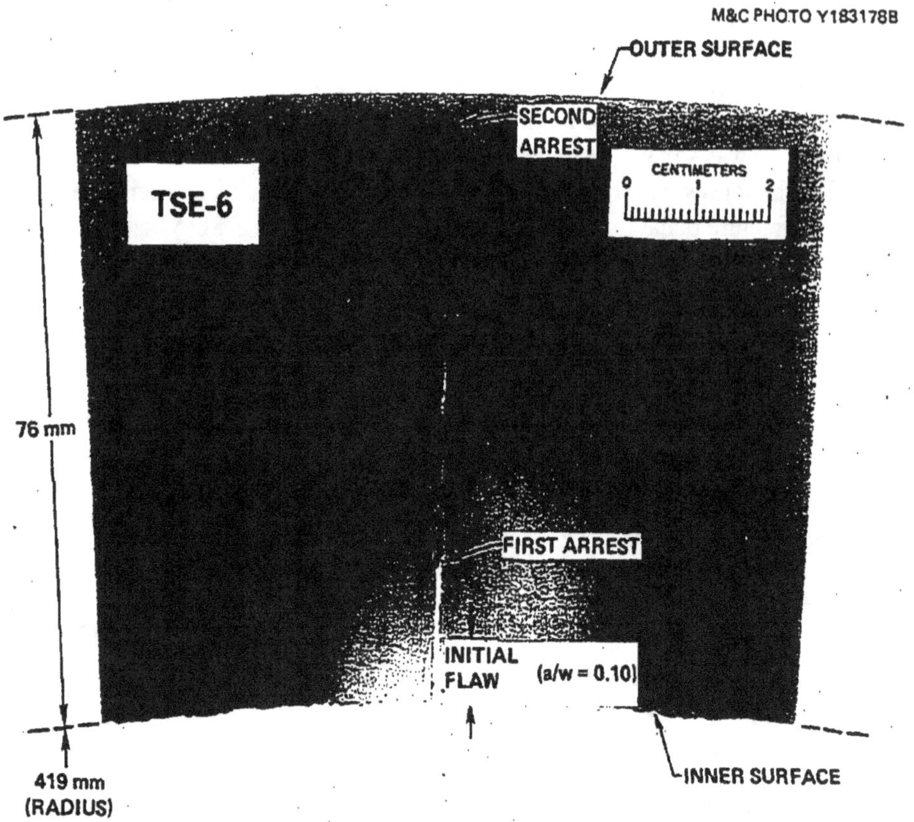

Fig. 10.4. Final radial profile of TSE-6 flaw near midlength of test cylinder.

that the longer crack jump and the less-stiff wall associated with TSE-6 contributed to this particular dynamic effect, which was a point of concern prior to TSE-6.

10.3 Crack Depths

As shown in Fig. 10.5, the initial flaw was uniform in depth (a/w = 0.10) and extended the full length of the test cylinder. Even so, the crack front associated with the first arrest event did not extend all the way to the ends, and the depth of the arrested front is somewhat less at midlength than elsewhere. Ignoring the tapering off at the ends of the flaw, the fractional depth (a/w) of the first arrested crack front varies from 0.21 to 0.32, and there is a 240-mm length in the lower half of the cylinder that has a uniform value of 0.28. With the exception of the tapering off at the ends, which is due to free-end effects, the variation of crack depth is probably due to small axial variations in fracture toughness.

The final crack depth was uniform over ~75% of the length of the test cylinder and was equal to 93% of the wall thickness (a/w = 0.93). At the ends of the cylinder, a/w = 0.82.

The crack depth associated with the very brief second arrest event was ~82% of the wall thickness (a/w = 0.82). Thus, this brief arrest event took place rather close to the site of the final arrest event, as was the case with TSE-5A.

10.4 Fracture-Mechanics Analysis

10.4.1 Posttest analysis based on design toughness curves

The pretest analysis for TSE-6 was based on toughness curves (TSE-6 design curves) deduced from the results of TSE-5 and indicated that only one initiation-arrest event would take place (see Fig. 4.9). Furthermore, a posttest analysis, based on actual temperatures and the design toughness curves, also indicates a single initiation-arrest event, as shown in Fig. 10.8. However, as discussed in Sect. 10.2, there were actually two "static" events and one "dynamic" event. As shown in Fig. 10.8, which includes the actual path of events, the first initiation-arrest event took place ~44 s earlier than "predicted," indicating that the actual fracture toughness was less than anticipated.

A quantitative comparison of K_{Ic} and K_{Ia} data can be obtained using Figs. 10.9 and 10.10, which correspond to Fig. 10.8 and are plots of T, K_I, K_{Ic}, and K_{Ia} vs a/w for the two times when events took place. These figures show that K_{Ic} and K_{Ia} for the first initiation-arrest event were ~40 and ~14% less, respectively, than the design values. For the second initiation and final arrest event, K_{Ic} and K_{Ia} were somewhat greater (24 and 18%, respectively) than the design values.

10.4.2. K_{Ic} and K_{Ia} values deduced from TSE-6

The calculated critical values of K_I corresponding to the first two initiation events and the first and last arrest events are listed in Table 10.1, which also includes for these events the times, crack depths, and crack-tip temperatures. These critical values of K_I are compared with the TSE-6 design curves and the lab small-specimen data in Fig. 10.11. As indicated in this figure, the values of K_{Ic} and K_{Ia} deduced from TSE-6 agree reasonably well with the curves corresponding to the final set of small-specimen data, although the K_{Ic} and K_{Ia} values corresponding to the second initiation event and the final arrest event seem somewhat high.

10.4.3 Posttest analysis based on final set of lab small-specimen K_J and K_{Ia} data

The few TSE-6 data points in Table 10.1 and Fig. 10.11 could not be used in a meaningful way to construct K_{Ic} and K_{Ia} curves for a final posttest analysis of TSE-6. Instead, a final analysis was performed

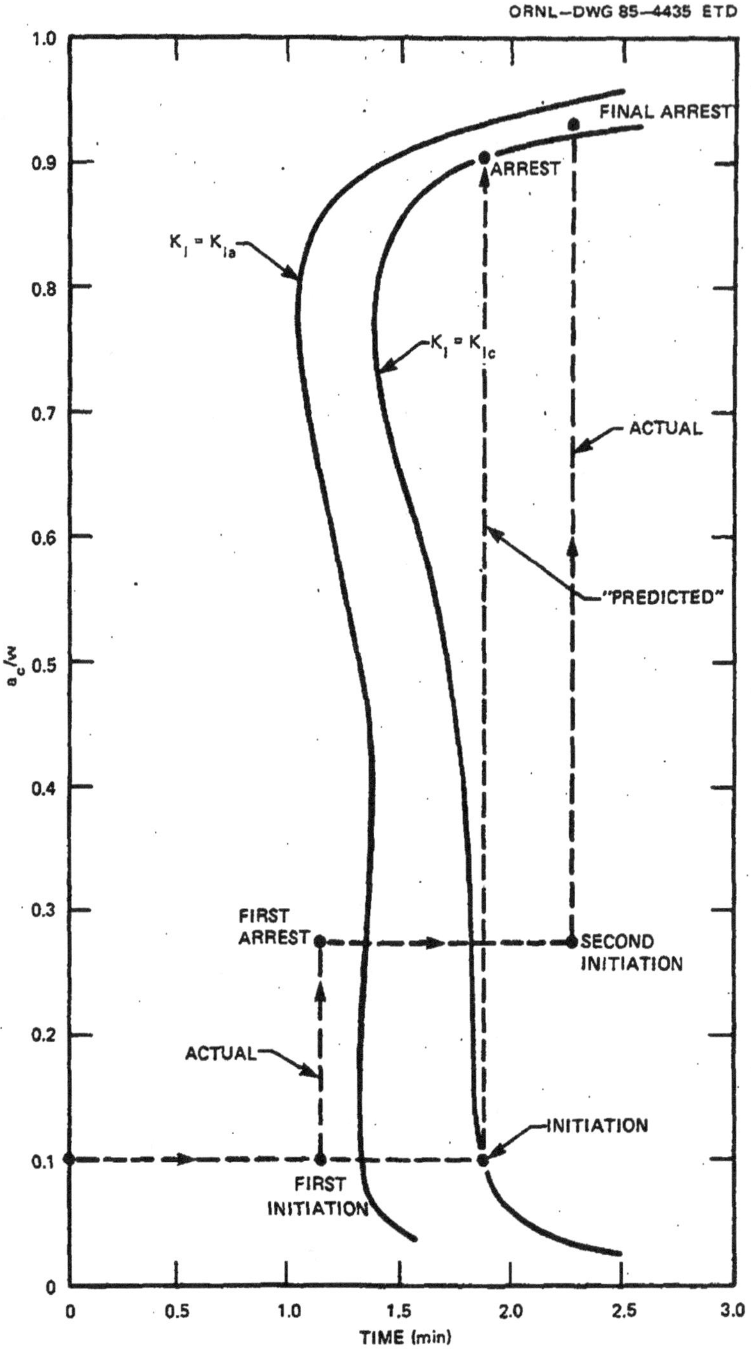

Fig. 10.8. TSE-6 posttest critical-crack-depth curves, based on measured temperatures and design toughness curves, with actual events superimposed.

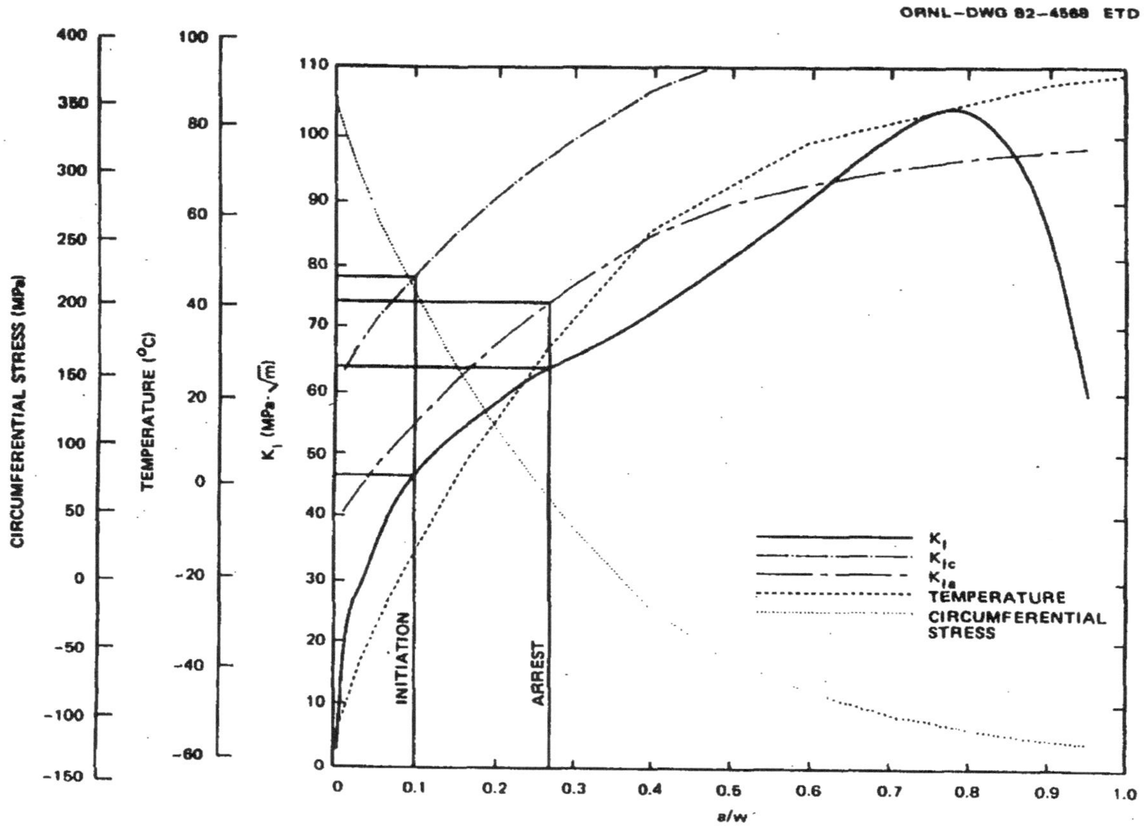

Fig. 10.9. Characteristic fracture-mechanics curves for TSE-6 at the time of the first initiation-arrest event (69 s), based on posttest analysis using measured temperatures and design toughness curves.

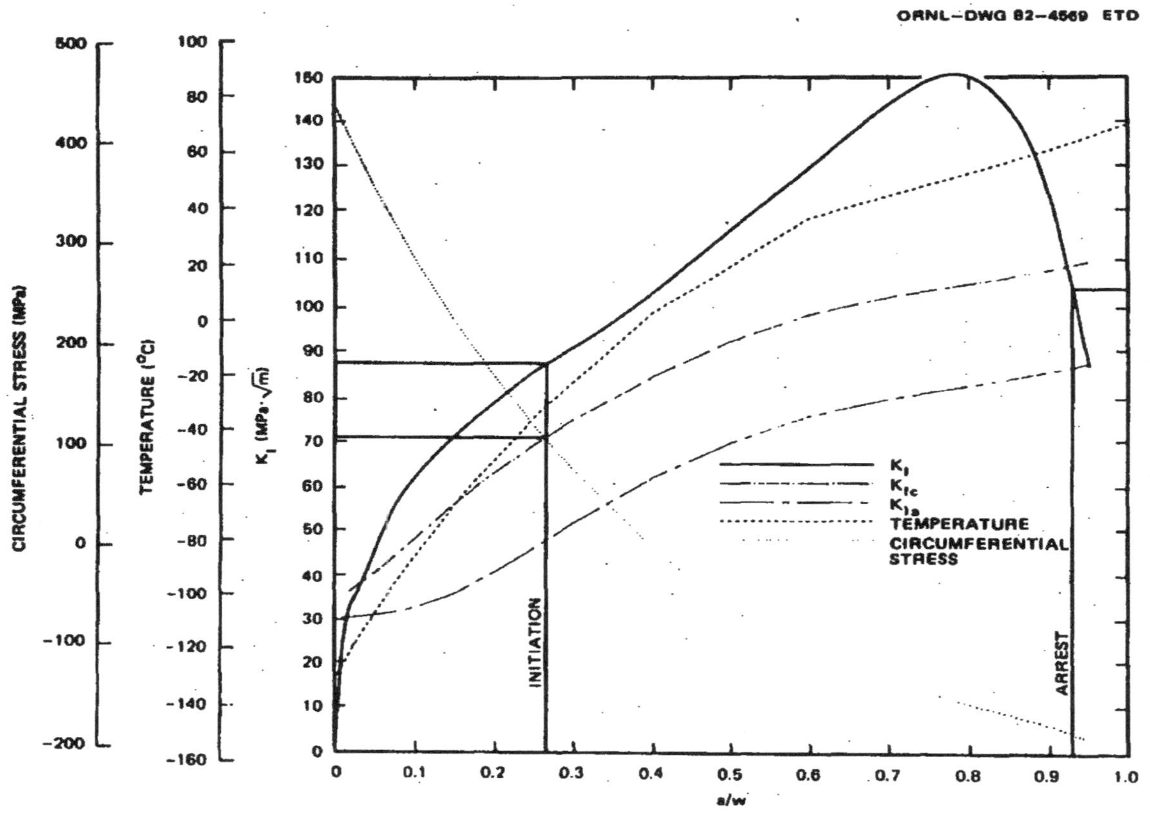

Fig. 10.10. Characteristic fracture-mechanics curves for TSE-6 at the time of the second initiation-arrest event (137 s), based on posttest analysis using measured temperatures and design toughness curves.

Table 10.1. Summary of events for TSE-6

Time (s)	Event	a/w	K_I (MPa·$\sqrt{m}$)	Temperature (°C)
69	Initiation	0.10	46	−13
	Arrest	0.28	63	32
137	Initiation	0.28[a]	87	−31
	Arrest	0.93[b]	105	63

[a]Second initiation event.

[b]Final arrest event.

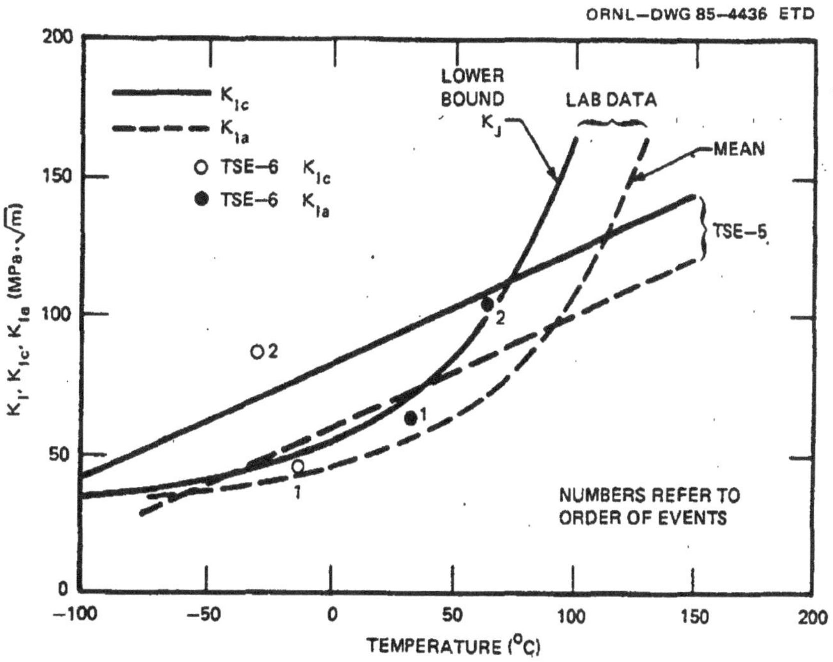

Fig. 10.11. Comparison of TSE-6 K_{Ic} and K_{Ia} values, TSE-6 K_{Ic} and K_{Ia} design curves (deduced from TSE-5 results), and small-specimen K_J, K_{Ic}, and K_{Ia} data.

using the small-specimen lower-bound K_J and the mean K_{Ia} curves shown in Figs. 7.7 and 10.11. The results are shown in Fig. 10.12, which is a set of critical-crack-depth curves with the actual events superimposed, and in Figs. 10.13 and 10.14, which are plots of temperature, K_I, K_{Ic} and K_{Ia} vs a/w for the two times at which events took place. (A complete set of digital output is included in Appendix F.) These results indicate reasonably good agreement between experiment and analysis for the first

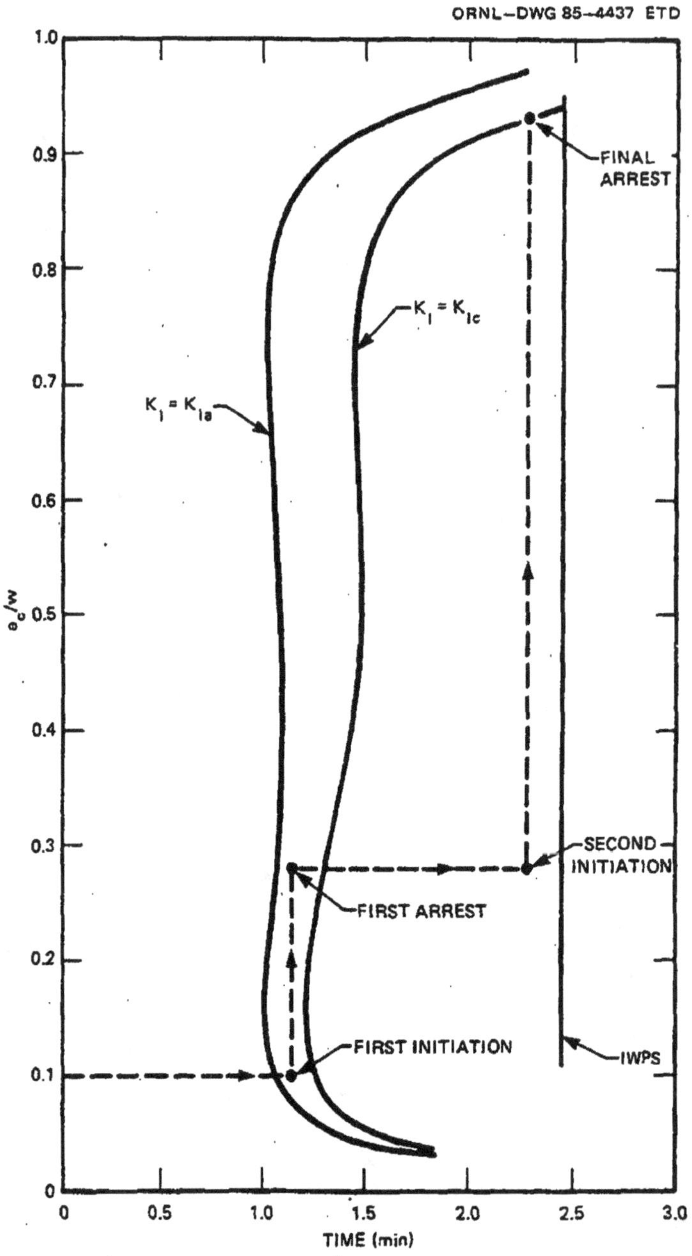

Fig. 10.12. TSE-6 posttest critical-crack-depth curves, based on measured temperatures and lab small-specimen lower-bound K_J and mean K_{Ia} data, with actual events superimposed.

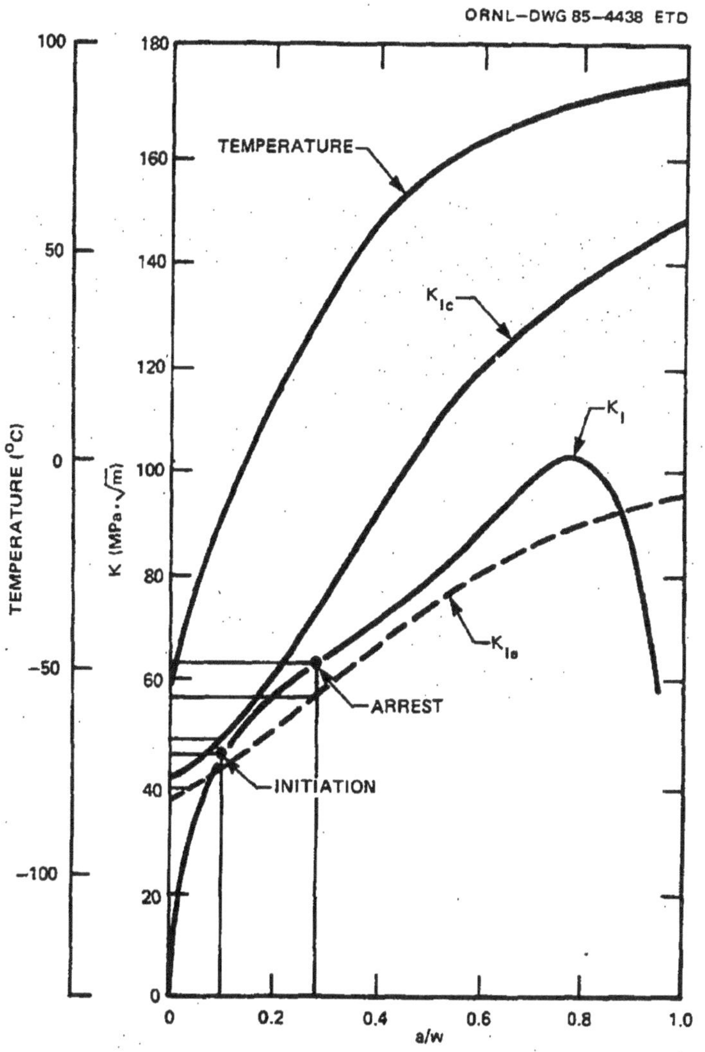

Fig. 10.13. Characteristic fracture-mechanics curves for TSE-6 at time of first initiation-arrest event (69 s), based on posttest analysis using small-specimen fracture-toughness data.

initiation, first arrest, and final arrest event and a rather large discrepancy for the second initiation event, consistent with the data in Fig. 10.11. The critical values of K_I corresponding to the first initiation and arrest events are ~6% less and ~12% greater, respectively, than the corresponding toughness curves. For the second initiation event, the critical value of K_I is ~88% greater than the initiation toughness curve.

A comparison of K_I and K_{Ia} for the final arrest event may not be very meaningful because of the proximity of the crack tip to the outer surface of the cylinder and because of the apparent very steep gradient in K_I, which makes the calculated critical value of K_I very sensitive to the effective crack depth. Probably the most important observation is

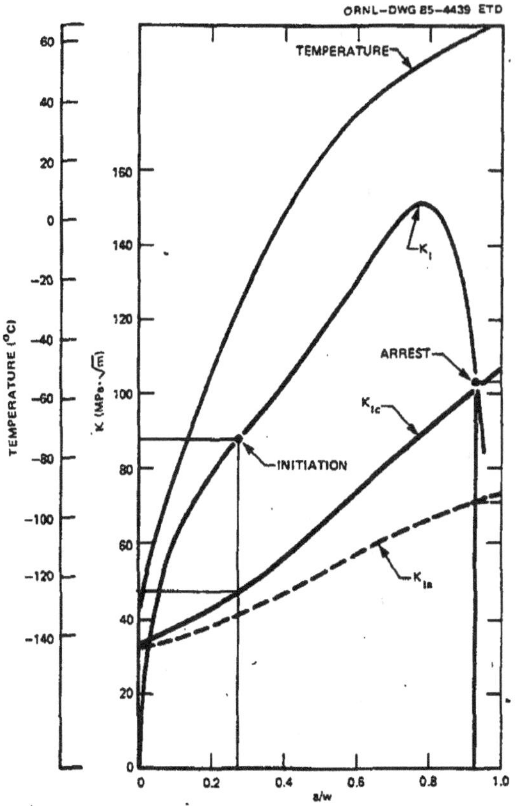

Fig. 10.14. Characteristic fracture-mechanics curves for TSE-6 at time of second initiation-arrest event (137 s), based on posttest analysis using small-specimen fracture-toughness data.

that almost irrespective of the value of K_{Ia}, the calculation would predict arrest deep in the wall (a/w > 0.9) and, indeed, that is what happened.

2.4.4. Crack Initiation and Arrest Values from TSE-5, 5A, and 6

(Cheverton 85a and 86) compared the critical values of K_I corresponding to crack initiation and arrest events with laboratory specimen data (Figs. 3 and 4 below).

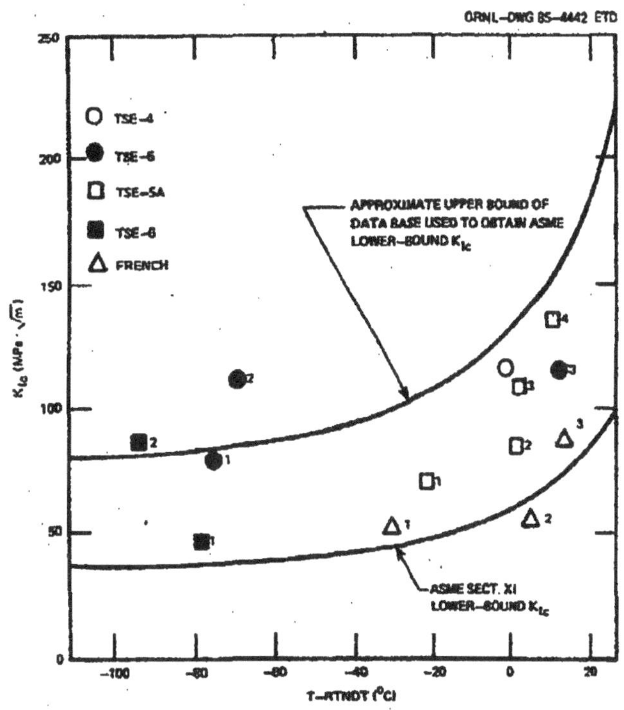

Fig. 3. Comparison of K_{Ic} data from TSE cylinder tests and laboratory specimens.

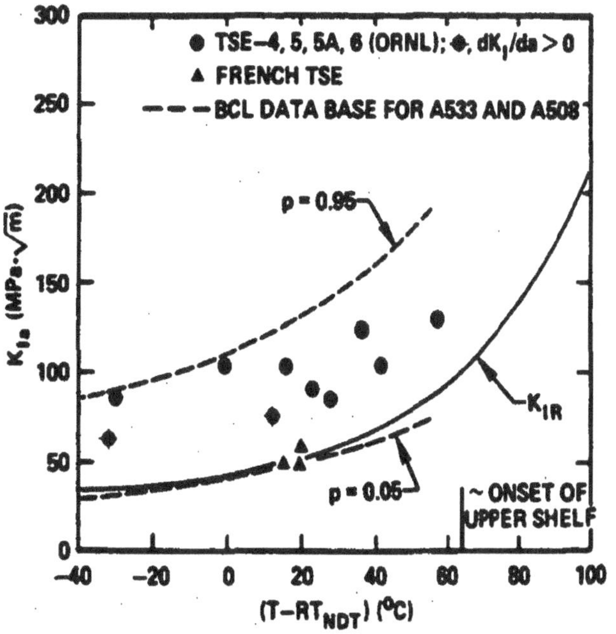

Fig. 4. Comparison of K_{Ia} data from TSE cylinder tests and laboratory specimens.

The curves shown in these figures are the upper- and lower-bound curves from small-specimen data. Overall, the K_{Ic} and K_{Ia} values derived from these TSE experiments demonstrate that the fracture behavior under these large-scale thermal-shock situations are adequately predicted by the use of LEFM methods and fracture properties obtained from tests of small laboratory specimens.

3. Summary Discussion

The discussion in this Appendix has shown that multiple flaw initiation-arrest events are credible for thick-wall cylinders exposed to thermal-shock transients. Additionally, it has shown that the nature and extent of such fracture behavior can be adequately predicted by careful application of linear elastic fracture mechanics analyses. This discussion has centered on the series of Thermal-Shock Experiments that was conducted within the HSST program at ORNL. Because the basic factors driving the fracture behavior in these TSEs are so similar to those for PTS scenarios, multiple fracture run-arrest events are deemed credible for an RPV exposed to PTS transient loads.

Appendix F – Detailed Reply to Peer Review Comment #75

Reviewer Murley commented that *"to better understand the PFM results it would be useful to see examples where the progress of a crack through the vessel wall is tracked."* In this Appendix we discuss four transients for which we track the progress (or lack thereof) of various simulated cracks through the vessel wall. One transient was selected from each of the four dominant transient classes:

- Primary side pipe break
  - Beaver Valley transient 07
  - 8 in. surge line break
- Stuck-open valve on the primary side
  - Oconee transient 122
  - Stuck-open pressurizer safety valve that recloses at 6000 seconds. Operator throttles HPI 10 minutes after reaching the throttling criteria
- Main steam line break
  - Beaver Valley transient 104
  - Main steam line break with AFW continuing to feed affected generator for 30 minutes. Operator controls HHSI 60 minutes after allowed. Break is assumed to occur inside containment so that the operator trips the RCPs due to adverse containment conditions
- Stuck-open valve on the secondary side
  - Palisades transient 55
  - Turbine/reactor trip with 2 stuck-open ADVs on SG-A combined with controller failure resulting in the flow from two AFW pumps into affected steam generator. Operator starts second AFW pump.

We selected flaws to track to illustrate the various features of the FAVOR crack initiation/arrest/reinitiat/rearrest model. In FAVOR flaws can initiate (start moving through the vessel wall), arrest (stop moving through the vessel wall), and ultimately fail the vessel in a number of different ways:

- Cracks can **initiate from original fabrication flaws** only by cleavage fracture (i.e., K_I exceeds K_{Ic}). Note that the criteria for cleavage crack initiation also requires that K_I be rising when it exceeds K_{Ic}. If K_I is falling when it exceeds K_{Ic} a condition exists called "warm pre-stress" and crack initiation can no longer occur. In principal crack initiation by ductile mechanisms is also possible (i.e., K_I exceeds K_{JIc}). However for the combinations of flaw sizes, loadings, and toughness conditions considered in this project initiation from an original fabrication flaw by ductile mechanisms has never been simulated.
- Cracks **arrest** whenever the driving force (K_I) falls below the cleavage crack arrest toughness (K_{Ia}).
- Once arrested, cracks can **reinitiate** by one of two mechanisms
  - Cracks can reinitiate in cleavage (see description under *initiate from original fabrication flaws* above).
  - Cracks can reinitiate by ductile tearing if the applied driving force exceeds the ductile crack initiation toughness (i.e., if K_I exceeds K_{JIc}).

  Note that the mode of crack reinitiat is controlled by the lesser of the cleavage crack initiation toughness and the ductile crack initiation toughness.

- Once reinitiated, cracks can **rearrest** by one of two mechanisms
    - Cracks can rearrest in cleavage (see description under *arrest* above).
    - Cracks can rearrest due to inadequate driving force to continue propagation of a ductile crack.

  Note that the mode of crack rearrest is controlled by the mode of crack reinitiat.
- The crack initiation/arrest/reinitiat/rearrest process continues until either a stable arrest is achieved somewhere in the vessel wall or the vessel is simulated to fail. **Through-wall cracking (failure) of the vessel** can occur by any of the following three mechanisms:
    - Net-section collapse of the ligament between the crack tip and the vessel OD can occur (tensile instability).
    - Ductile tearing can become structurally unstable
    - The crack can advance by either cleavage or ductile mechanisms to a user-specified fraction of the total wall thickness. In the analyses reported herein that fraction is set to 90%. Accurate solution of the fracture driving force equations for very deep cracks is not possible, necessitating use of this cut-off value.

Details of all of the fracture models can be found in both [EricksonKirk 10-03] and in [Williams 10-03].

The remaining pages of this appendix track the progress (or lack thereof) of various simulated cracks through the vessel wall for the transients described earlier.

Deterministic analysis of a simulated flaw subjected to Beaver Valley transient 07 at 60 EFPY

Transient sequence:	Beaver Valley 07	
Transient Description:	8 in. surge line break (See Figure 7.1 for pressure and temperature variation)	
Flaw Analyzed:	Orientation:	Axial
	Type:	Embedded flaw in plate material
	Depth (2a):	0.321 in
	Length (2c):	0.446 in
	Inner crack tip distance from ID ($\mathscr{L}$):	0.218 in
Embrittlement:	EFPY:	60 years
	Simulated RT_{NDT} at inner crack tip	279°F

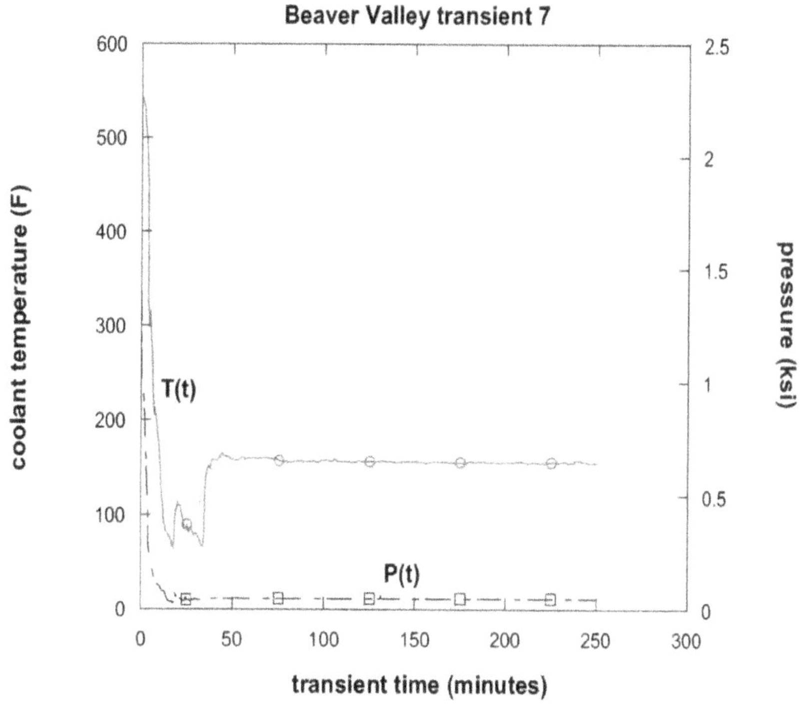

Figure 7.1 – Beaver Valley transient sequence 7 – 8 in. surge line break

Crack Initiation

Figure 7.2 illustrates that the conditions for the flaw to have a conditional probability of crack initiation (cpi) > 0 are satisfied: (1) the applied driving force to fracture (K_I) is greater than the minimum of the cleavage fracture initiation toughness (K_{Ic}) distribution (which corresponds to the Weibull 'a' parameter), and (2) during the time that applied K_I is greater than the minimum K_{Ic}, the applied K_I must also be greater than at all previous time steps. The second condition is a necessary condition to overcome effect of warm-prestress.

Figure 7.2 illustrates that the flaw has a conditional probability of initiation (cpi) > 0 in the transient time interval between 10 and 12 minutes. The flaw cannot initiate before a transient time of 10 minutes since this is the first time step at which K_I exceeds the minimum value of K_{Ic}. The flaw cannot initiate after a transient time of 12 minutes because this is the time at which the maximum applied K_I occurs, producing a condition of warm-prestress.

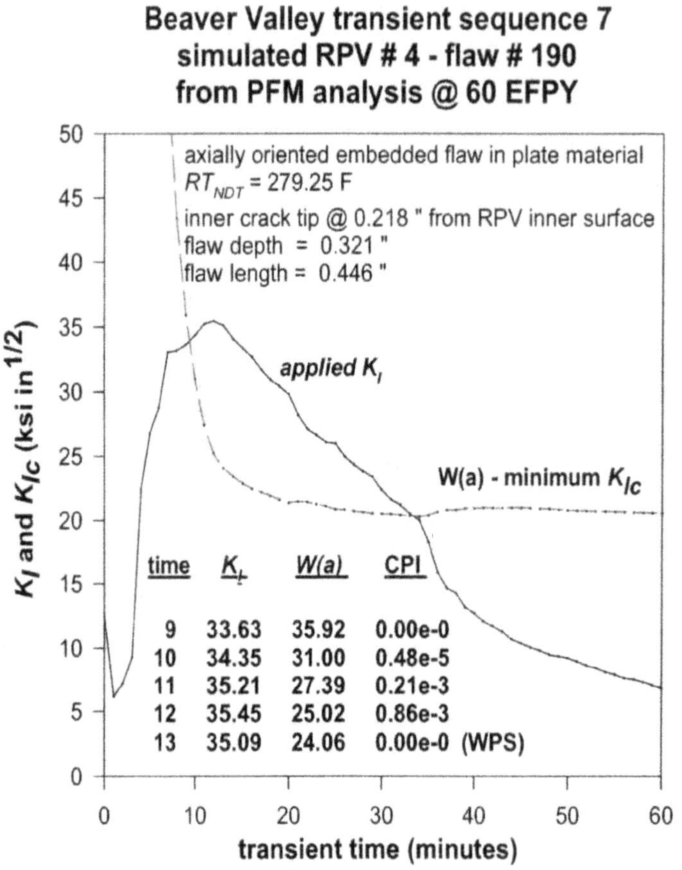

Figure 7.2 – Beaver Valley transient sequence 7 – deterministic LEFM analysis for flaw from PFM Monte Carlo analysis for which cpi > 0

Through-Wall Cracking Analysis #1:

This is one simulation of the through-wall cracking behavior of the flaw initiated in Figure 7.2. This analysis occurs at t=12 minutes.

Event 1: Figure 7.3(a) illustrates that the initiated flaw propagates through the wall thickness to failure since the applied driving force to fracture (K_I) exceeds the crack arrest toughness (K_{Ia}) through the entire wall thickness at t=12 minutes.

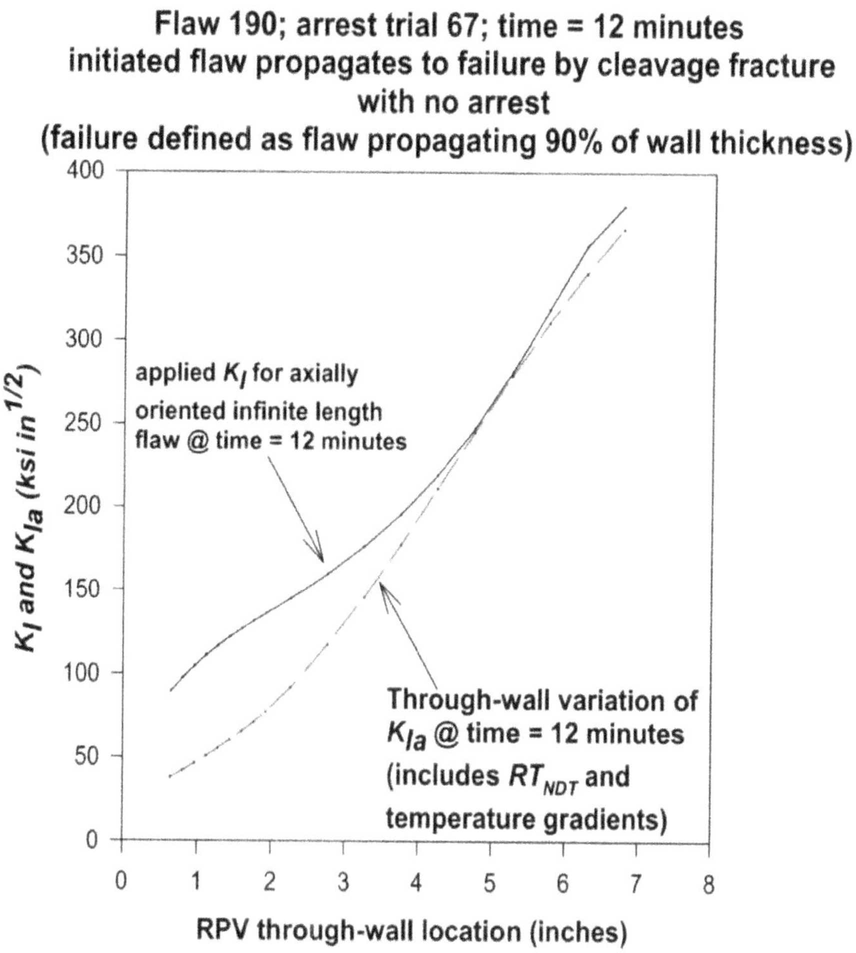

Figure 7.3 – Beaver Valley transient sequence 7 – deterministic through-wall analysis for flaw initiated at t = 12 (illustrated in figure 7.2). Vessel is considered as failed since flaw propagated 90% of the distance through the wall.

Through-Wall Cracking Analysis #2:

This is a different simulation of through-wall cracking behavior of the flaw that initiated in Figure 7.2. The simulation has a progression different from the first simulation because of different sampled values for the cleavage and ductile crack initiation toughness values. This analysis is performed at t=12 minutes.

Event 1: Figure 7.4(a) illustrates that the initiated flaw propagates to a depth of 1.77 in. where it arrests since the applied driving force to fracture (K_I) falls below the crack arrest toughness (K_{Ia}).

Event 2: Figure 7.4(b) illustrates that the arrested flaw does not reinitiate in cleavage fracture since the applied driving force to fracture (K_I) does not exceed the cleavage fracture initiation toughness (K_{Ic}). Nor does the flaw reinitiate by ductile tearing since the applied driving force to fracture (K_I) does not exceed the upper shelf crack initiation fracture toughness (K_{JIc}). This flaw has experienced a stable arrest and does not fail the vessel.

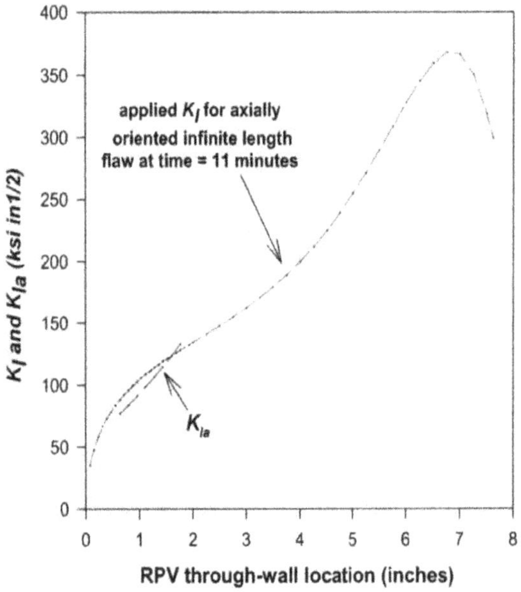

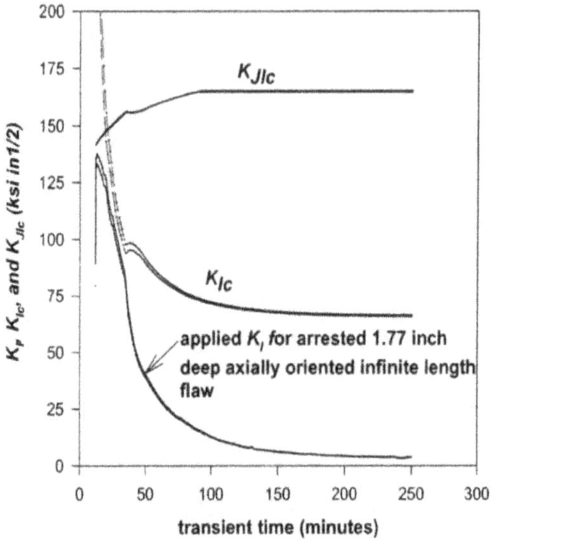

Figure 7.4(a) {left} – Beaver Valley transient sequence 7 – deterministic through-wall analysis for flaw initiated at t = 11 (illustrated in figure 7.2) for which case the flaw is arrested at a depth of 1.77 in. since $K_I < K_{Ia}$.

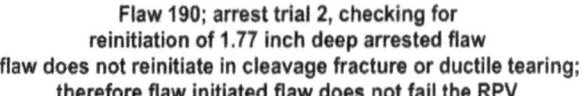

Figure 7.4(b) {right}– Beaver Valley transient sequence 7 – checking for re-initiation of arrested flaw illustrate in 7.4(a). Flaw does not reinitiate in cleavage fracture or ductile tearing.

Deterministic analysis of a simulated flaw subjected to Oconee transient 122 at 60 EFPY

Transient sequence:	Oconee 122	
Transient Description:	Stuck-open pressurizer safety valve that recloses at 6000 seconds (See Figure 122.1 for pressure and temperature variation)	
Flaw Analyzed:	Orientation:	Axial
	Type:	Embedded flaw in weld material
	Depth (2a):	0.604 in
	Length (2c):	0.966 in
	Inner crack tip distance from ID ($\mathcal{L}$):	0.854 in
Embrittlement:	EFPY:	60 years
	Simulated RT_{NDT} at inner crack tip	208°F

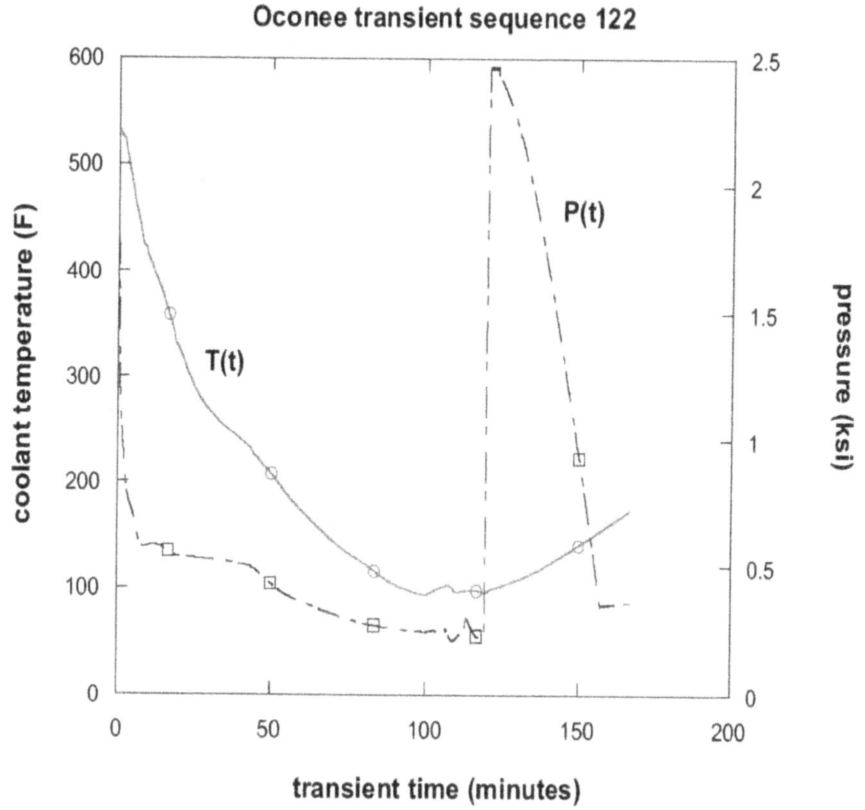

Figure 122.1 Oconee transient 122 – stuck-open pressurizer safety valve that recloses at 6000 seconds.

Crack Initiation

Figure 122.2 illustrates that the conditions for the flaw to have a conditional probability of crack initiation (cpi) > 0 are satisfied: (1) the applied driving force to fracture (K_I) is greater than the minimum of the cleavage fracture initiation toughness (K_{Ic}) distribution (which corresponds to the Weibull 'a' parameter), and (2) during the time that applied K_I is greater than the minimum K_{Ic}, the applied K_I must also be greater than at all previous time steps. The second condition is a necessary condition to overcome effect of warm-prestress.

Figure 122.2 illustrates that the flaw has a conditional probability of initiation (cpi) > 0 in the transient time interval between 119 and 121 minutes. The flaw cannot initiate before a transient time of 120 minutes since this is the first time step at which K_I exceeds the minimum value of K_{Ic}. The flaw cannot initiate after a transient time of 121 minutes because this is the time at which the maximum applied K_I occurs, so warm-prestress prevents crack initiation for all transient times > 121 minutes.

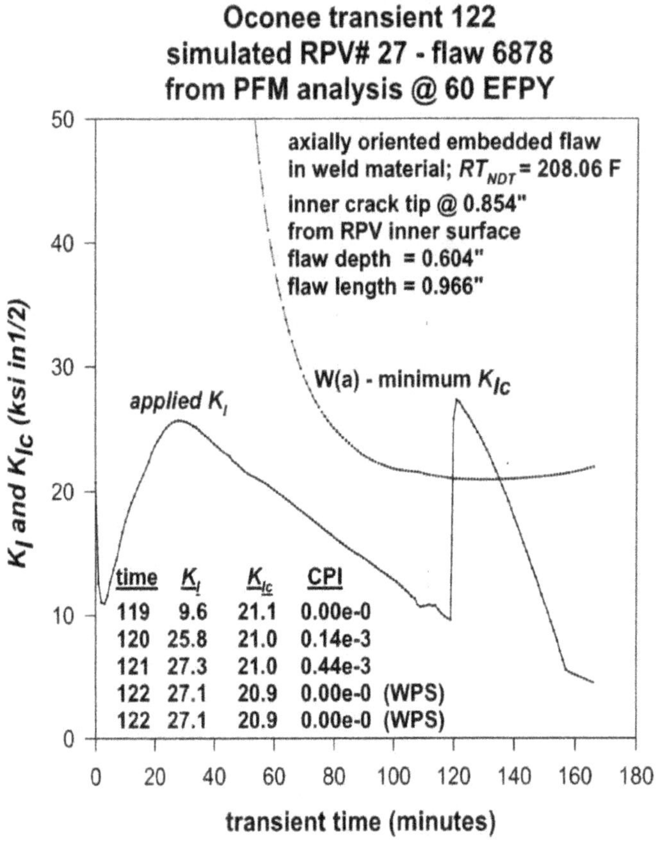

Figure 122.2 – Oconee transient 122 – deterministic LEFM analysis for flaw from PFM Monte Carlo analysis for which cpi > 0

Through-Wall Cracking Analysis #1:

This is one simulation of the through-wall cracking behavior of the flaw initiated in Figure 122.2. This analysis occurs at t=120 minutes.

Event 1: Figure 122.3 illustrates that the initiated flaw propagates through the wall thickness to failure since the applied driving force (K_I) exceeds the crack arrest toughness (K_{Ia}) through the entire wall thickness at t=120 minutes. The mode of failure is plastic instability.

Even though all of the other through-wall analyses performed at this time step have different sampled values for the cleavage and ductile crack initiation toughness values, they all fail due to plastic instability as illustrated in through-wall cracking analysis #1. For Oconee transient 122, all initiated flaws fail; therefore, the conditional probability of through-wall cracking is identical to the conditional probability of crack initiation.

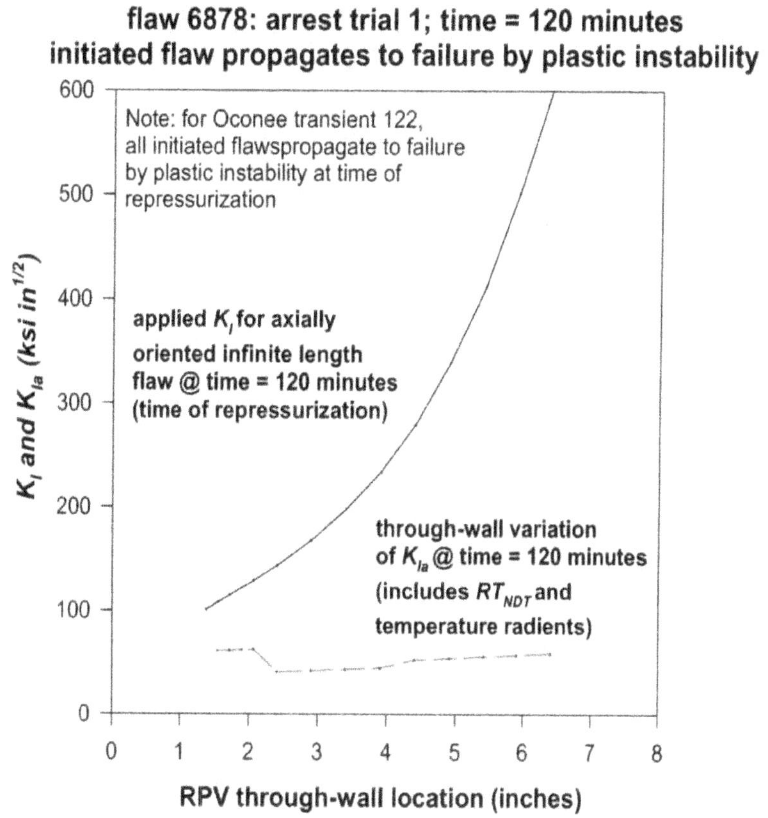

Figure 122.3 – Oconee transient 122 – deterministic through-wall analysis for flaw initiated at t = 120 minutes that results in failure by plastic instability

Deterministic analysis of a simulated flaw subjected to Beaver Valley transient 104 at 60 EFPY

Transient Sequence:	Beaver Valley 104	
Transient Description:	Main Steam Line Break with AFW continuing to feed affected generator for 30 minutes (See Figure 104.1 for pressure and temperature variation).	
Flaw Analyzed:	Orientation:	Circumferential
	Type:	Embedded flaw in weld material
	Depth (2a):	0.321 in
	Length (2c):	0.620 in
	Inner crack tip distance from ID ($\mathcal{L}$):	0.226 in
Embrittlement:	EFPY:	60 years
	Simulated RT_{NDT} at inner crack tip	319°F

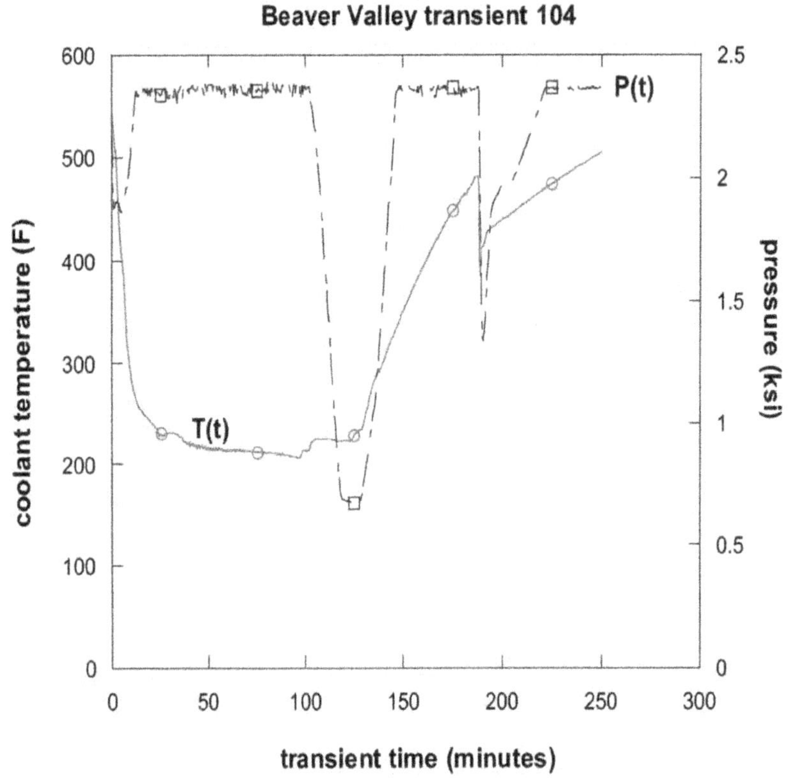

Figure 104.1 Beaver Valley transient sequence 104 – Main Steam Line Break with AFW continuing to feed affected generator for 30 minutes

Crack Initiation

Figure 104.2 illustrates that the conditions for the flaw to have a conditional probability of crack initiation (cpi) > 0 are satisfied: (1) (1) the applied driving force to fracture (K_I) is greater than the minimum of the cleavage fracture initiation toughness (K_{Ic}) distribution (which corresponds to the Weibull 'a' parameter), and (2) during the time that applied K_I is greater than the minimum K_{Ic}, the applied K_I must also be greater than at all previous time steps. The second condition is a necessary condition to overcome effect of warm-prestress.

Figure 104.2 illustrates that the flaw has a conditional probability of initiation (cpi) > 0 in the transient time interval between 11 and 12 minutes. The flaw cannot initiate before a transient time of 12 minutes since this is the first time step at which K_I exceeds the minimum value of K_{Ic}. The flaw cannot initiate after a transient time of 12 minutes because this is the time at which the maximum applied K_I occurs, so warm-prestress prevents crack initiation for all transient times > 12 minutes.

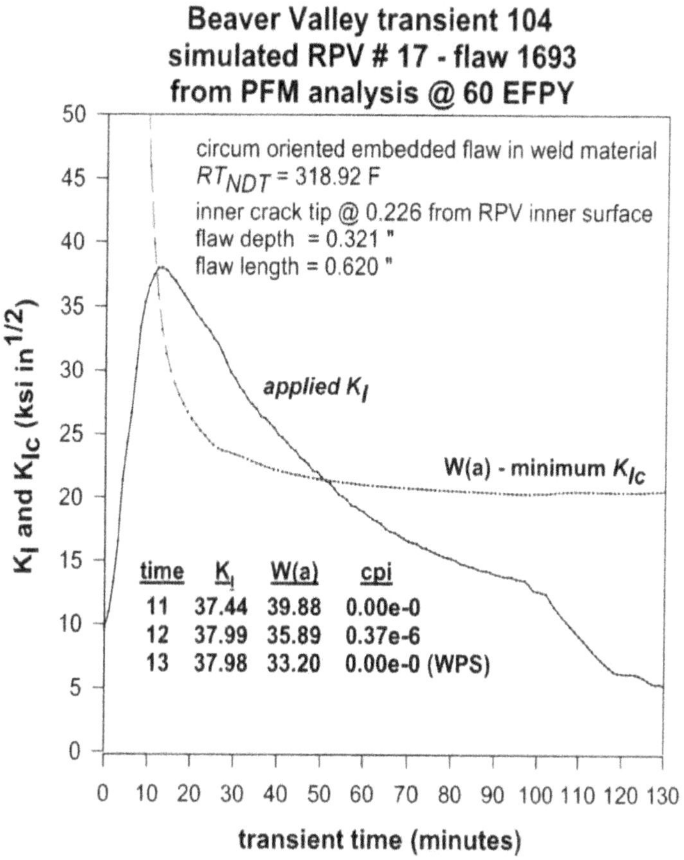

Figure 104.2 Beaver Valley transient sequence 104 – deterministic LEFM analysis of flaw from PFM Monte Carlo analysis for which cpi > 0

Through-Wall Cracking Analysis #1:

This is one simulation of the through-wall cracking behavior of the flaw initiated in Figure 104.2. This analysis begins at t=12 minutes.

Event 1: Figure 104.3(a) illustrates that the initiated flaw propagates to a depth of 1.93-in. where it arrests since the applied driving force to fracture (K_I) for the 360 degree continuous circumferential flaw falls below the crack arrest toughness (K_{Ia}).

Event 2: Figure 104.3(b) illustrates that the flaw arrested in figure 104.3(a) reinitiates at t=13 minutes by ductile tearing since the applied driving force to fracture (K_I) for the 360 degree continuous circumferential flaw is greater than the upper shelf crack initiation fracture toughness (K_{JIc}).

Event 3: Figure 104.3(c) illustrates that the flaw reinitiated by ductile tearing propagates by cleavage (since $K_I > K_{Ia}$) to a depth of 6.09 in. where it arrests since $K_I < K_{Ia}$. The FAVOR model allows a flaw that reinitiates by a stable ductile tear (of some finite distance) to resume cleavage fracture propagation if $K_I > K_{Ia}$. This is consistent with observations in large-scale fracture experiments [TSE REFS].

Event 4: Figure 104.3(d) illustrates that at t=14 minutes, the arrested flaw reinitiates in unstable ductile tearing; which propagates through the vessel wall, failing the vessel.

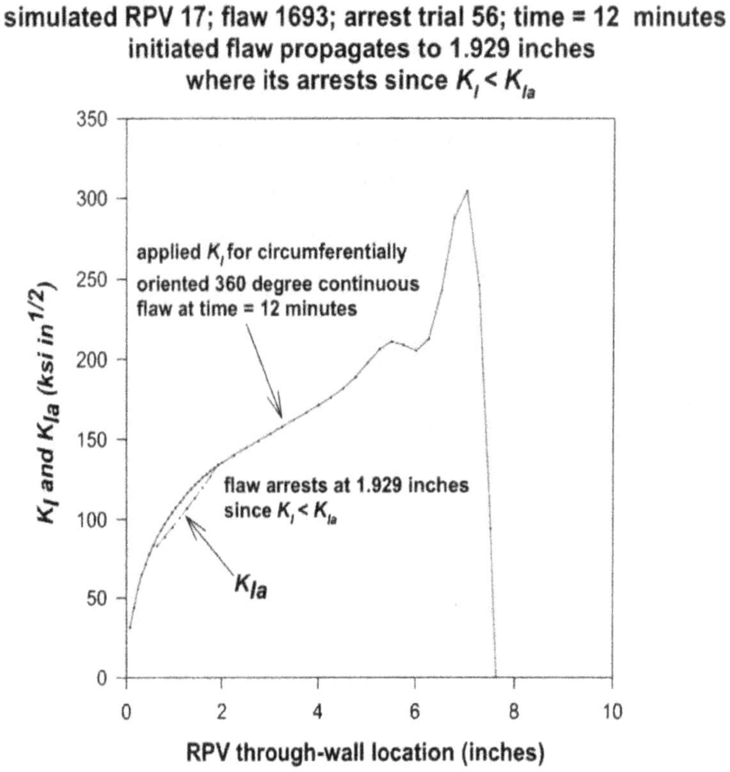

Figure 104.3(a) Beaver Valley transient sequence 104 – deterministic through-wall analysis for flaw initiated at t = 12 minutes (illustrated in figure 104.2) for which case the flaw is arrested since the applied driving force to fracture (K_I) falls below the crack arrest toughness (K_{Ia}).

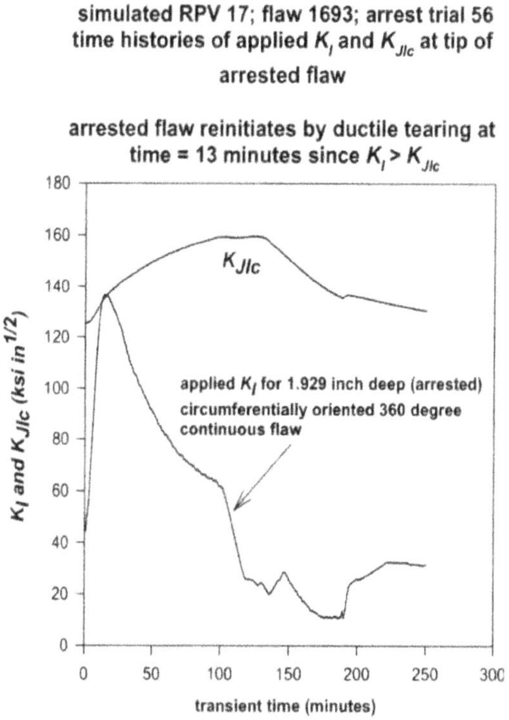

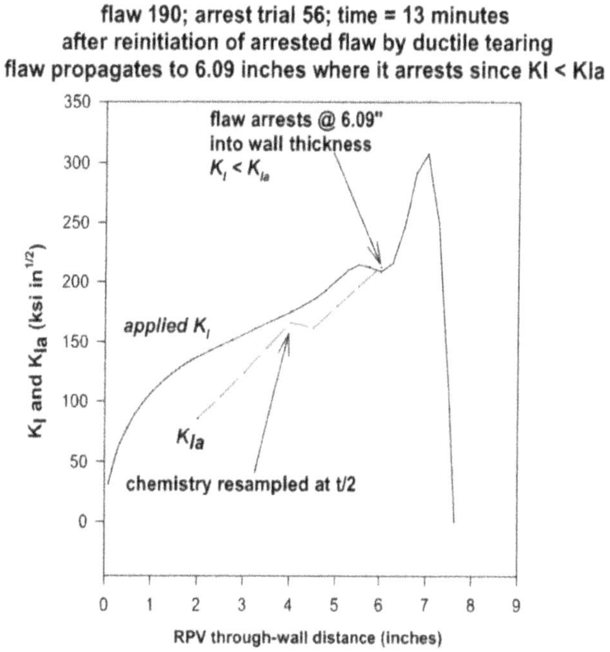

Figure 104.3(b) {left} Beaver Valley transient sequence 104 – deterministic analysis for arrested flaw illustrated in figure 104.3(a). Arrested flaw reinitiates by ductile tearing at time = 13 minutes since $K_I > K_{JIc}$.

Figure 104.3(c) {right} Beaver Valley transient sequence 104 – after re-initiation of arrested flaw by ductile tearing, flaw propagates to 6.09 in. where it arrests since $K_I < K_{Ia}$

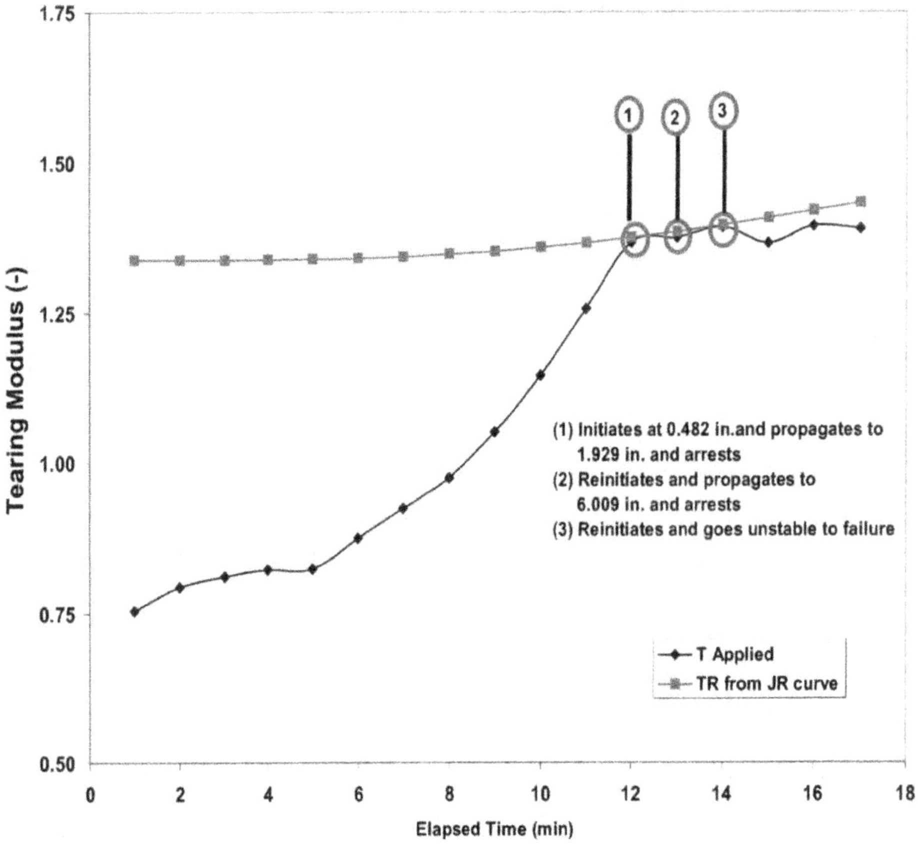

Figure 104.3(d) Beaver Valley transient sequence 104 –at time = 14 minutes, the 6.09 in. deep arrested flaw (illustrated in figure 104.3(c)) reinitiates in unstable ductile tearing to failure.

Through-Wall Cracking Analysis #2:

This is a different simulation of through-wall cracking behavior of the flaw that initiated in Figure 104.2. The simulation has a progression different from the first simulation because of different sampled values for the cleavage and ductile crack initiation toughness values. This analysis is performed at t=12 minutes.

Event 1: Figure 104.4(a) illustrates that the initiated flaw propagates to a depth of 1.92-in. where it arrests since the applied driving force to fracture (K_I) falls below the crack arrest toughness (K_{Ia}).

Event 2: Figure 104.4(b) illustrates that the arrested flaw does not reinitiate in cleavage fracture since the applied driving force to fracture (K_I) does not exceed cleavage fracture initiation toughness (K_{Ic}). Nor does the flaw reinitiate by ductile tearing since the applied driving force to fracture (K_I) does not exceed the upper shelf crack initiation fracture toughness (K_{JIc}). Therefore, this flaw has experienced a stable arrest and does not fail the vessel.

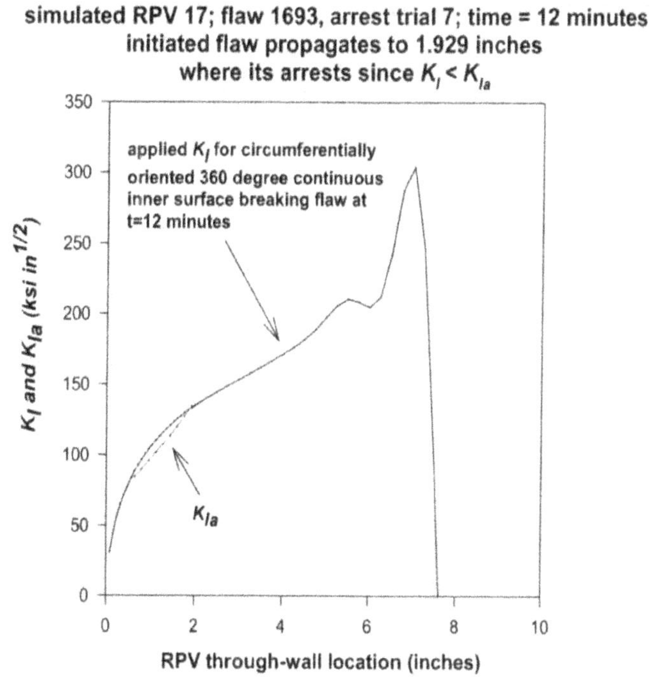

Figure 104.4(a) Beaver Valley transient sequence 104 – deterministic through-wall analysis for flaw initiated at t = 12 minutes (illustrated in figure 104.2) for which case the flaw is arrested.

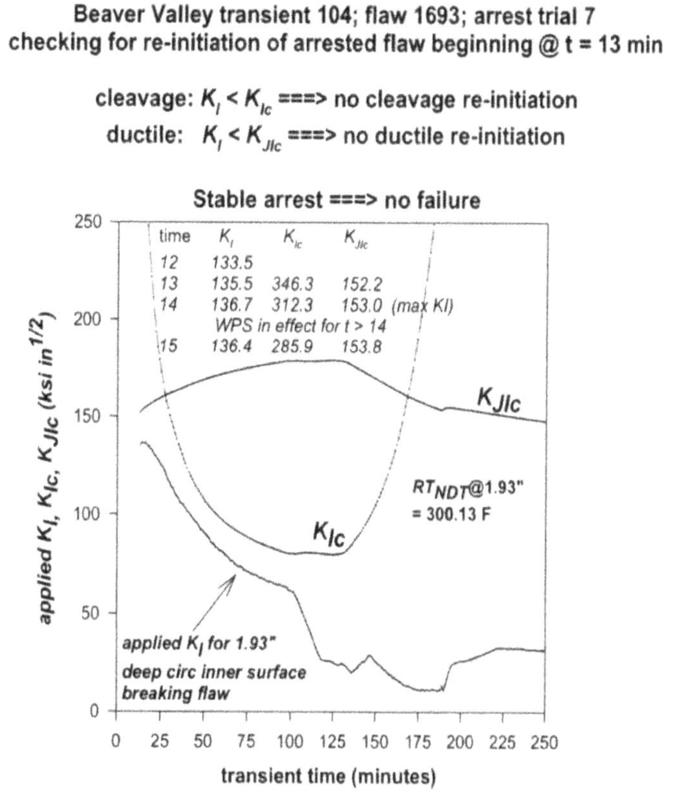

Figure 104.4(b) Beaver Valley transient sequence 104 – checking for re-initiation of arrested flaw illustrated in figure 104.4 (a). Flaw does not reinitiate in cleavage fracture or ductile tearing; therefore initiated flaw does not propagate to failure.

Deterministic analysis of a simulated flaw subjected Palisades transient 55 at 60 EFPY

Transient sequence:	Palisades transient 55	
Transient Description:	Turbine/reactor trip with two stuck-open valves combined with controller failure (See Figure 55.1 for pressure and temperature variation)	
Flaw Analyzed:	Orientation:	Axial
	Type:	Embedded flaw in weld material
	Depth (2a):	0.263 in
	Length (2c):	0.928 in
	Inner crack tip distance from ID ($\mathcal{L}$):	0.342 in
Embrittlement:	EFPY:	60 years
	Simulated RT_{NDT} at inner crack tip	390°F

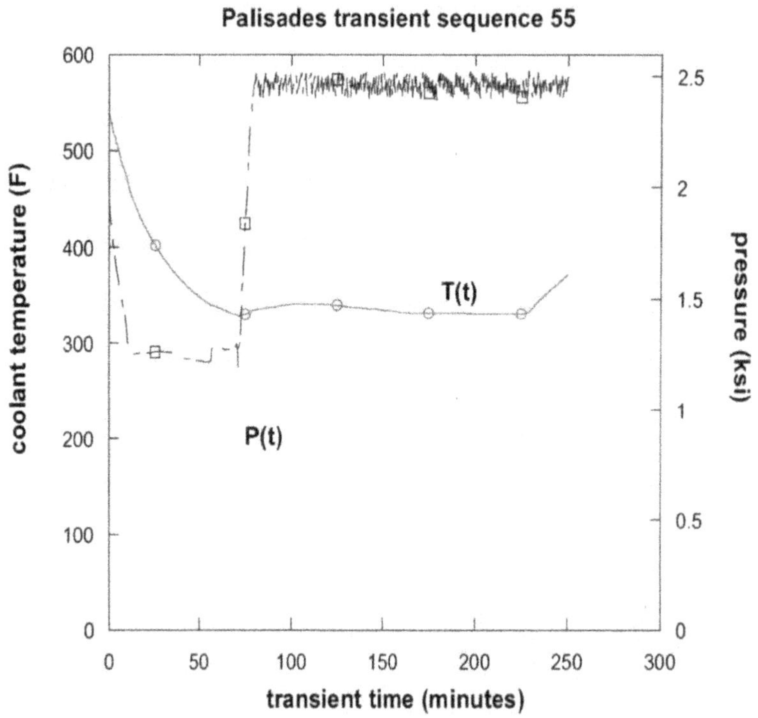

Figure 55-1 Palisades transient sequence 55 that results from a turbine/reactor trip with two stuck-open valves combined with controller failure.

Crack Initiation

Figure 55.2 illustrates that the conditions for the flaw to have a conditional probability of crack initiation (cpi) > 0 are satisfied: (1) the applied driving force to fracture (K_I) is greater than the minimum of the cleavage fracture initiation toughness (K_{Ic}) distribution (which corresponds to the Weibull 'a' parameter), and (2) during the time that applied K_I is greater than the minimum K_{Ic}, the applied K_I must also be greater than at all previous time steps. The second condition is a necessary condition to overcome effect of warm-prestress.

Figure 55.2 illustrates that the flaw has a conditional probability of initiation (cpi) > 0 in the transient time interval between 78 and 80 minutes. The flaw cannot initiate before a transient time of 78 minutes since this is the first time step at which K_I exceeds the minimum value of K_{Ic}. The flaw cannot initiate after a transient time of 80 minutes because this is the time at which the maximum applied K_I occurs, so warm-prestress prevents crack initiation for all transient times > 80 minutes.

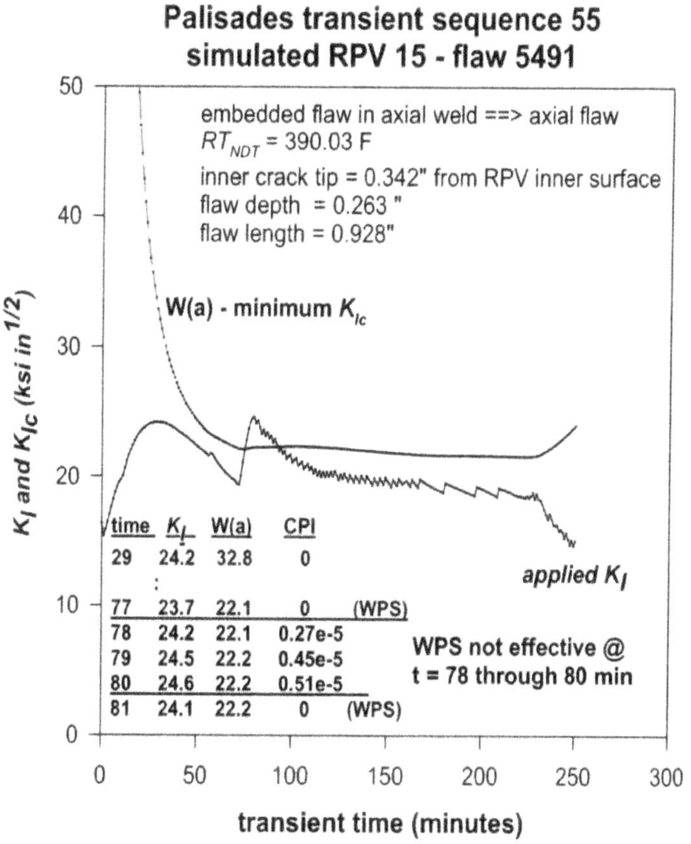

Figure 55-2 Palisades transient sequence 55 – deterministic LEFM analysis for flaw from PFM Monte Carlo analysis for which cpi > 0

Through-Wall Cracking Analysis #1:

This is one simulation of the through-wall cracking behavior of the flaw initiated in Figure 55.2. This analysis occurs at t=78 minutes.

Event 1: Figure 55.3 illustrates a deterministic through-wall analysis at time = 78 minutes in which the initiated flaw propagates through the wall, since the applied driving force to fracture (K_I) exceeds the crack arrest toughness (K_{Ia}), to a depth such that the failure is by plastic instability. It should be noted that in this case failure by plastic instability occurred at a more shallow depth that propagation to 90% of the wall thickness. Had the failure not occurred by plastic instability, from figure 55.3, it is clear that flaw would have propagated to 90% of the wall thickness in cleavage and therefore would have been considered as failed. Failure by plastic instability is a common mode of failure associated with transients that have repressurizations.

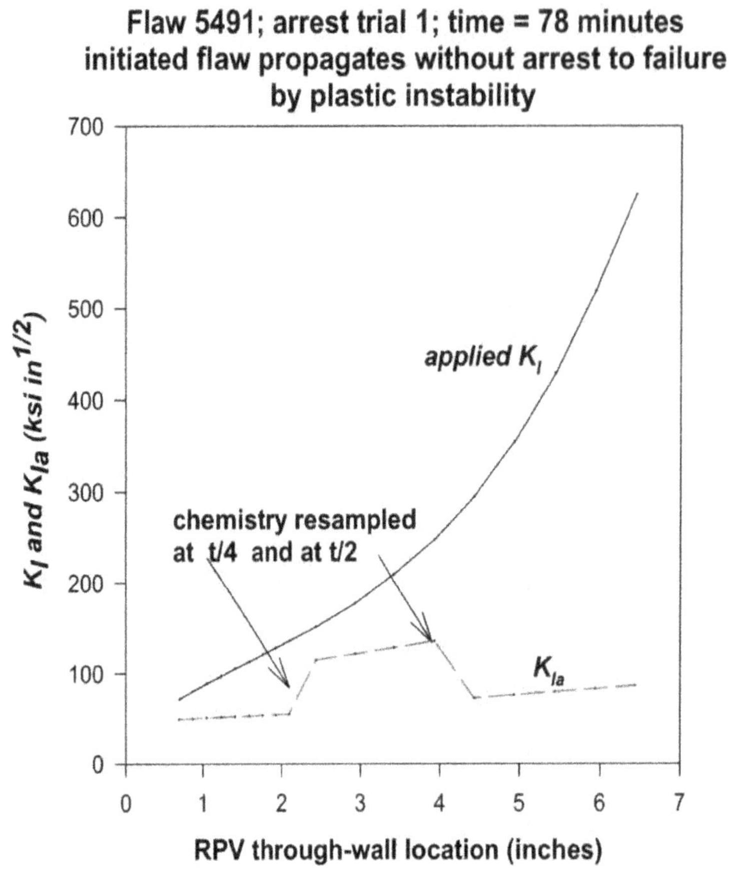

Figure 55-3 Palisades transient sequence 55 – deterministic through-wall analysis for flaw initiated at 78 minutes that propagates through the wall without arrest resulting in failure by plastic instability.

Through-Wall Cracking Analysis #2:

This is a different simulation of through-wall cracking behavior of the flaw that initiated in Figure 55.2. The simulation has a progression different from the first simulation because of different sampled values for the cleavage and ductile crack initiation toughness values. This analysis is also performed at t=78 minutes.

Event 1: Figure 55.4(a) illustrates that the initiated flaw propagates to a depth of 2.44 in. where it arrests since the applied driving force to fracture (K_I) falls below the crack arrest toughness (K_{Ia}). Note that the discontinuity in crack arrest toughness is due to a re-sampling of chemistry, which is performed at t/4, t/2, and 3t/4 through-wall locations for weld material.

Event 2: Figure 54.4(b) illustrates that the flaw arrested in figure 55.4(a) reinitiates at t=79 minutes by ductile tearing since the applied driving force to fracture (K_I) for the infinite length inner-surface breaking axially oriented flaw is greater than the upper shelf crack initiation fracture toughness (K_{JIc}) and that the applied K_I for the 2.44 in. deep flaw is greater than at previous time steps.

Event 3: Figure 54.4(c) illustrates that the flaw reinitiated by ductile tearing propagates by cleavage (since $K_I > K_{Ia}$) at time=79 minutes to a depth of 2.69 in. where it arrests since $K_I < K_{Ia}$.

Event 4: Figure 54.4(d) illustrates checking for re-initiation of the arrested flaw beginning a time=80 minutes. The arrested flaw does not reinitiate in cleavage fracture since the applied driving force to fracture (K_I) does not exceed cleavage fracture initiation toughness (K_{Ic}). Nor does the flaw reinitiate by ductile tearing, even though the applied driving force to fracture (K_I) does exceed the upper shelf crack initiation fracture toughness (K_{JIc}), however, at times after the maximum value of the applied driving force to fracture (K_I), therefore, warm prestress inhibits re-initiation by ductile tearing.

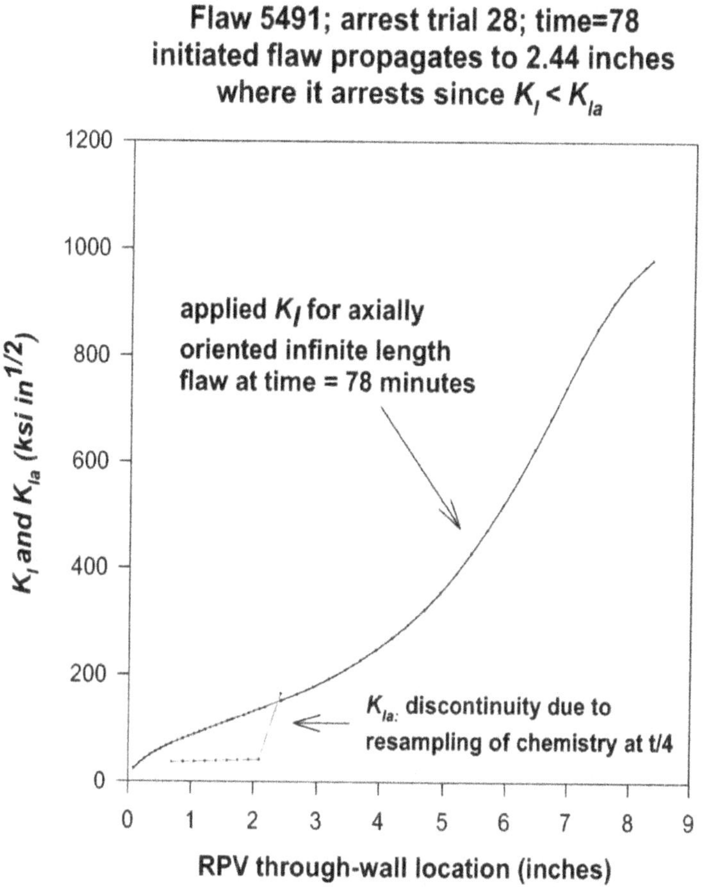

Figure 55-4(a) Palisades transient sequence 55 – deterministic through-wall analysis for flaw initiated at 78 minutes (illustrated in figure 55-2) that propagates to 2.44 where it arrest since $K_I < K_{Ia}$. The discontinuity in K_{Ia} is due to re-sampling of chemistry at t/4.

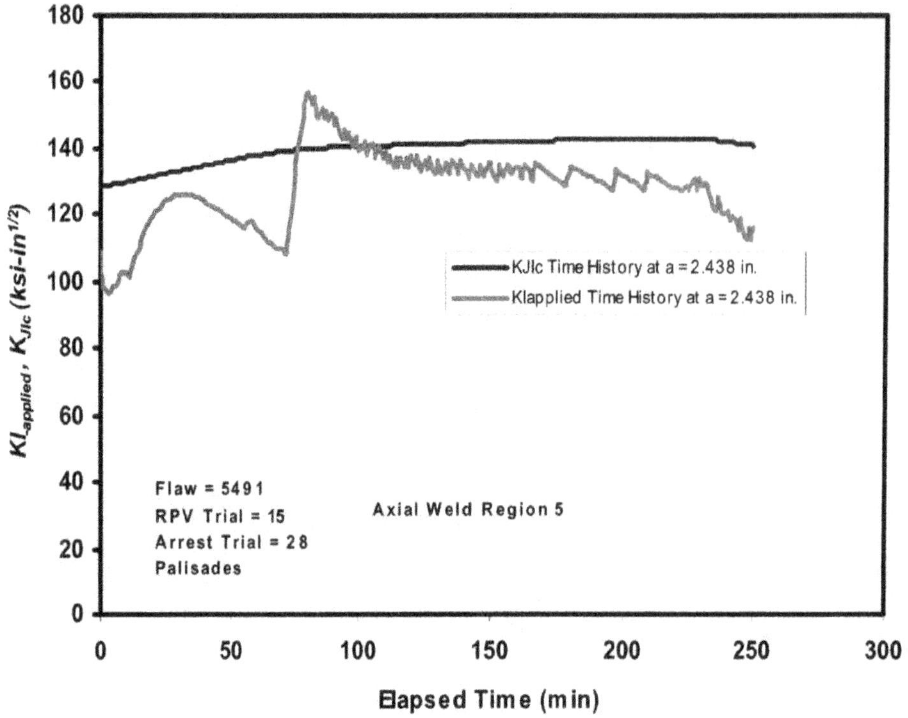

Figure 55-4(b) Palisades transient sequence 55 – arrested flaw illustrated in figure 55-4(a) reinitiates in ductile tearing at time = 79 since $K_I > K_{JIc}$

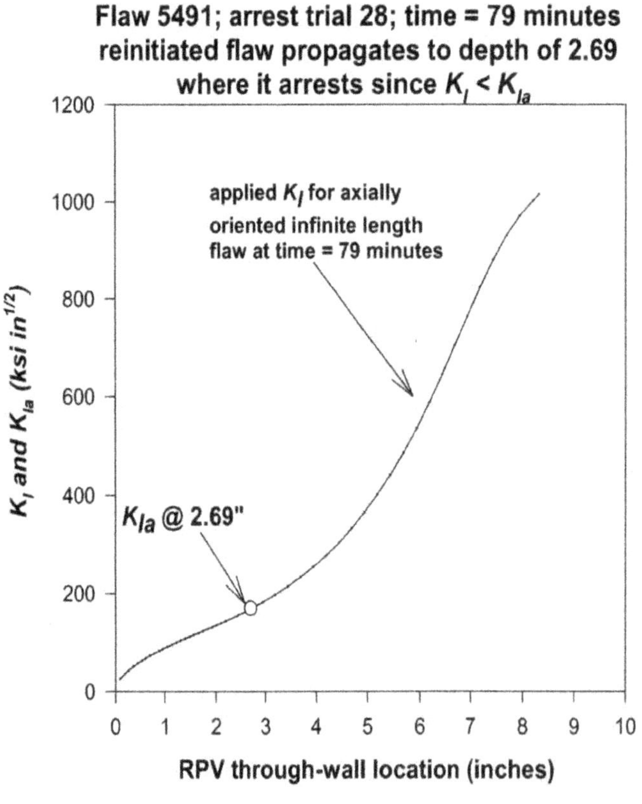

Figure 55-4(c) Palisades transient sequence 55 – 2.44 in. flaw that reinitiated in ductile tearing propagates to depth 2.69 in where it arrests since $K_I < K_{Ia}$

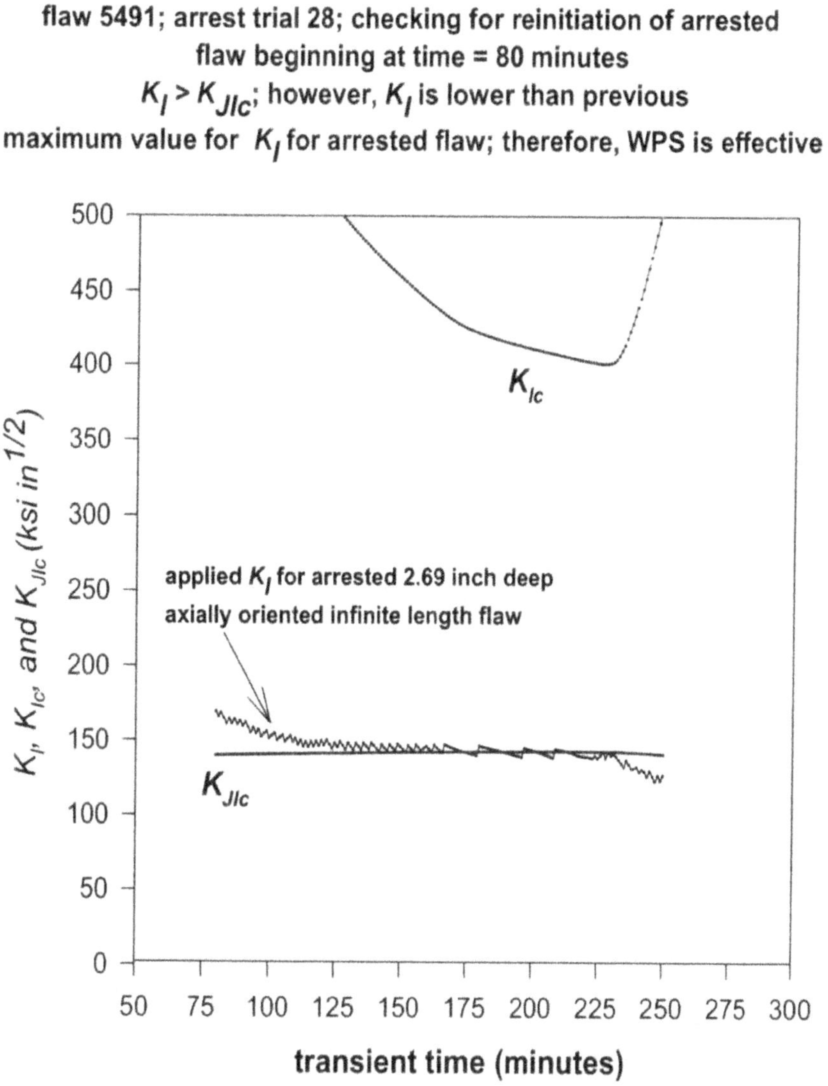

Figure 55-4(d) Palisades transient sequence 55 –checking for re-initiation of arrested flaw illustrated in figure 55-4(c). Flaw does not reinitiate in cleavage fracture ($K_I < K_{Ic}$) or ductile tearing. $K_I > K_{JIc}$; however, at times later than the maximum K_I occurred for arrested flaw; therefore WPS inhibits re-initiation by ductile tearing.

Appendix G – Flaw Distributions for Forgings

This Appendix includes two articles prepared by Dr. Frederic Simonen of the Pacific Northwest National Laboratory concerning flaw distributions in forgings. The staff used these articles as the basis of the forging flaw sensitivity studies reported in Section 9.2.2.2.

Technical Basis for the Input Files to FAVOR Code for Flaws in Vessel Forgings

F.A. Simonen
Pacific Northwest National Laboratory
Richland, Washington

July 28, 2004

Pacific Northwest National Laboratory (PNNL) has been funded by the U.S. Nuclear Regulatory Commission (NRC) to generate data on fabrication flaws that exist in reactor pressure vessels (RPV). Work has focused on flaws in welds but with some attention also to flaws in the base metal regions. Data from vessel examinations along with insights from an expert judgment elicitation (MEB-00-01) and from applications of the PRODIGAL flaw simulation model (NUREG/CR-5505, Chapman et. al. 1998) have been used to generate input files (see report NUREG/CR-6817, Simonen et. al. 2003) for probabilistic fracture mechanics calculations performed with the FAVOR code by Oak Ridge National Laboratory. NUREG/CR-6817 addresses only flaws in plate materials and provided no guidance for estimating the numbers and sizes of flaws in forging materials. More recent studies have examined forging material, which has provided a data on flaws that were detected and sized in the examined material. At the request of NRC staff PNNL has used these more recent data to supplement insights from the expert judgment elicitation to generate FAVOR code input files for forging flaws. The discussion below describes the technical basis and results for the forging flaw model.

Nature of Base Metal Flaws

PNNL examined material from some forging material from a Midland vessel as described by Schuster (2002). The forging was made during 1969 by Ladish. Examined material included only part of the forging that had been removed from the top of the forged ring as scrap not intended for the vessel. This material was expected to have more than the average flaw density, and as such may contribute to the conservatism of any derived flaw distribution.

Figures 1 and 2 show micrographs of small flaws in plate and forging materials. These flaws are inclusions rather than porosity or voids. They are also not are planer cracks. Therefore their categorization as simple planar or as volumetric flaws is subject to judgment. The plate flaw of Figure 1 has many sharp and crack-like features, whereas such features are not readily identified for the particular forging flaw seen in Figure 2. It should, however, be emphasized that the PNNL examined only a limited volume of both plate and forging material and found very few flaws in examined material. It is not possible to generalize from such a small sample of flaws. Accordingly, the flaw model makes assumptions that may be somewhat conservative, due to the limited data on the flaw characteristics.

Flaw Model for Forging Flaws

The model for generating distributions of forging flaws for the FAVOR code uses the same approach as that for modeling plate flaws as described in NUREG/CR-6817. The quantitative results of the expert elicitation are used along with available data from observed forging flaws. The flaw data were used as a "sanity check" on the results of the expert elicitation. Figure 3 summarizes results of the expert elicitation. Each expert was asked to estimate ratios between flaw densities in base metal compared to the corresponding flaw densities observed in the weld metal of the PVRUF vessel. Separate ratios were requested for plate material and forging material.

As indicated in Figure 3, the parameters for forging flaws are similar to those for plate flaws. The forging and plate models used the same factor of 0.1 for the density of "small" flaws (flaws with through-wall dimensions less than the weld bead size of the PVRUF vessel). The density of "large" flaws in forging material is somewhat greater than the density of flaws in plate material. The factor of 0.025 for the flaw density is replaced by a factor of 0.07 for forging flaws. A truncation level of 0.11 mm is used for both plate and forging flaws. As described in the next section the data from forging examinations show that these factors are consistent with the available data. It is noted that the assumption for the 0.07 factor is supported by only a single data point corresponding to the largest observed forging flaw (with a depth dimension of 4 mm).

The factors of 0.1 and 0.07 came from the recommendations from the expert elicitation on vessel flaws. As noted below the very limited data from PNNL's examinations of forging material show that these factors are consistent with the data, although the 0.07 factor is supported by only one data point for an observed forging flaw with a 4-mm depth dimension.

Comparison with Data on Observed Flaws

The PNNL examinations of vessel materials included both plate materials and forging materials. For plate flaws less than 4-mm in through-wall depth dimension, Figure 4 shows data from NUREG/CR-6817 that shows frequencies for plate flaws. Also shown for comparison are the flaw frequencies for the welds of the PVRUF and Shoreham vessels. This plot confirmed results of the expert judgment elicitation (Figure 4) and indicated: 1) there are fewer flaws in plate material than in weld material, and 2) there is about a 10:1 difference in flaw frequencies for plates versus welds.

PNNL generated the data on flaws in forgings after preparation of NUREG/CR-6817. Forging data are presented in Figures 5 and 6 along with the previous data for flaws in the PVRUF plate material. There is qualitative agreement with the results of the expert judgment elicitation (Figure 4), which indicates that 1) plate and forging materials have similar frequencies for small (2 mm) flaws, and 2) forging material have higher flaw frequencies for larger (>4 mm) flaws.

Inputs for FAVOR Code

Figure 7 compares the flaw frequencies for plates and forgings that were provided to ORNL as input files for the FAVOR code. This plot shows mean frequencies from an uncertainty distribution as described by the flaw input files. It is seen that the curves for plate and forging flaws are identical for small flaws but show differences for the flaws larger than 3% of the vessel wall thickness. Also seen is the effect of truncating the flaw distribution at a depth of 11 mm (about 5% of the wall thickness).

References

Jackson, D.A. and L. Abramson, 2000. *Report on the Preliminary Results of the Expert Judgment Process for the Development of a Methodology for a Generalized Flaw Size and Density Distribution for Domestic Reactor Pressure Vessel*, MED-00-01, PRAB-00-01,U.S. Nuclear Regulatory Commission.

Schuster, G.J., 2002. "Technical Letter Report – JCN-Y6604 – Validated Flaw Density and Distribution Within Reactor Pressure Vessel Base Metal Forged Rings," prepared by Pacific Northwest National Laboratory for U.S. Nuclear Regulatory Commission, December 20, 2002.

Simonen, F.A., S.R.Doctor, G.J. Schuster and P.G. Heasler, 2003. *A Generalized Procedure for Generating Flaw-Related Inputs for the FAVOR Code*, NUREG/CR-6817, Rev. 1, prepared by Pacific Northwest National Laboratory for U.S. Nuclear Regulatory Commission.

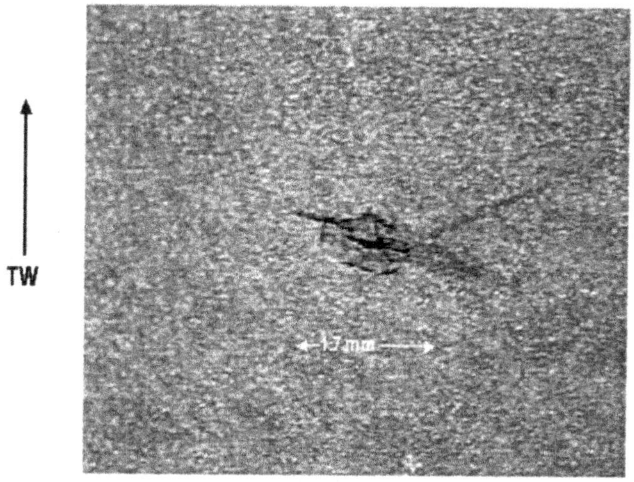

Figure 1 Small Flaw in Plate Material

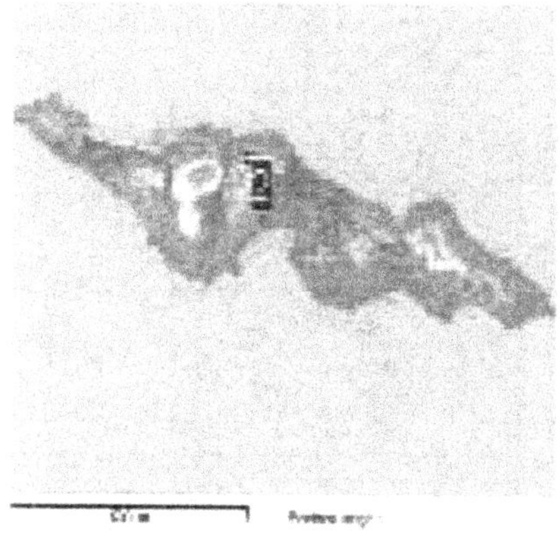

Figure 2 Small Flaw in Forging Material

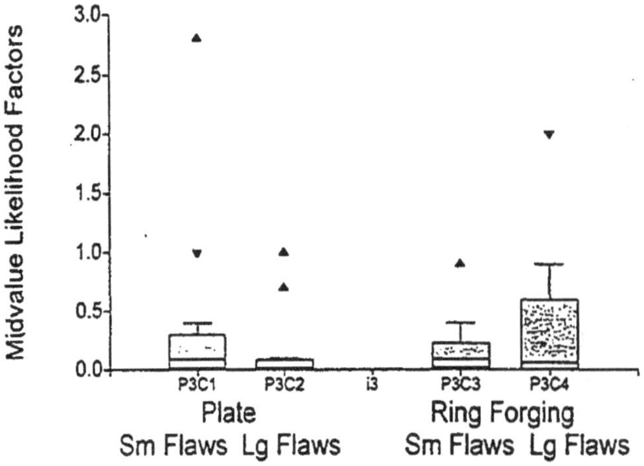

	Base Metal vs. Weldmetal			
	Plate vs. Welds		Ring Forgings vs. Welds	
	Small Flaws	Large Flaws	Small Flaws	Large Flaws
MIN	.0004	.001	.001	.002
LQ	.015	.01	.02	.007
MED	.1	.025	.1	.07
UQ	.3	.09	.2	.6
MAX	12.0	1.0	.9	2.0

Figure 3 Relative Flaw Densities of Base Metal Compared to Weld Metal as Estimated by Expert Judgment Process (from Jackson and Abramson 2000)

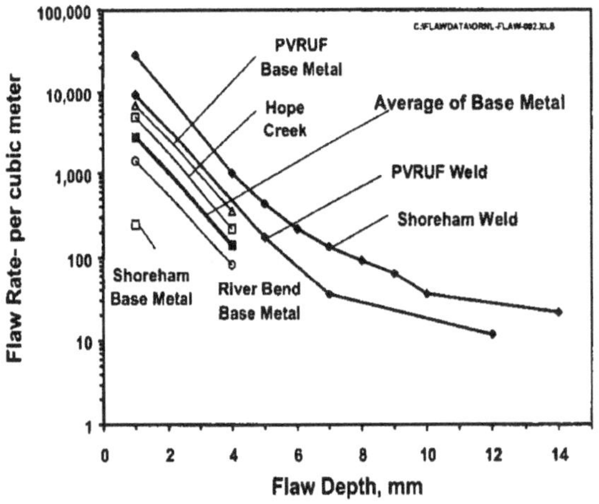

Figure 4. Flaw Frequencies for Plate Materials with Comparisons to Data for Weld Flaws

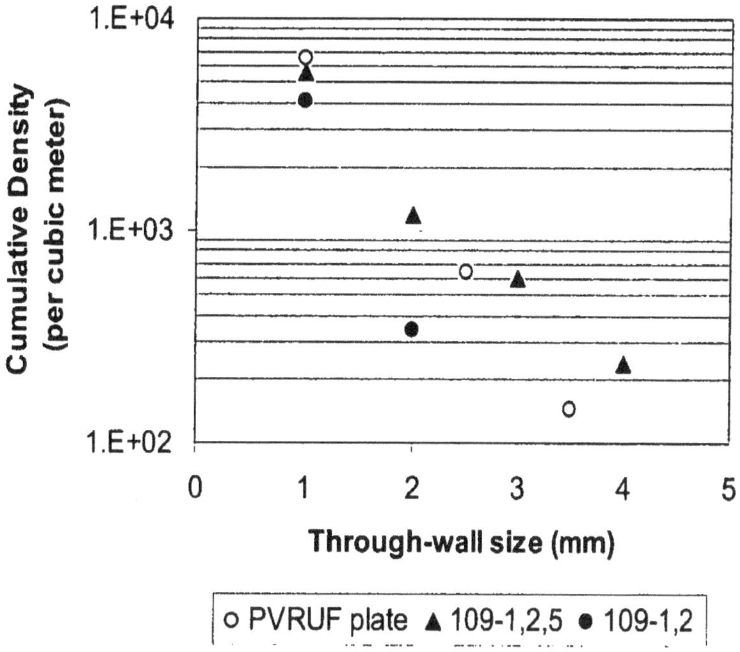

Figure 5. **Validated Flaw Density and Size Distribution for Three Forging Specimens. Cumulative flaw density is the number of flaws per cubic meter of equal or greater size.**

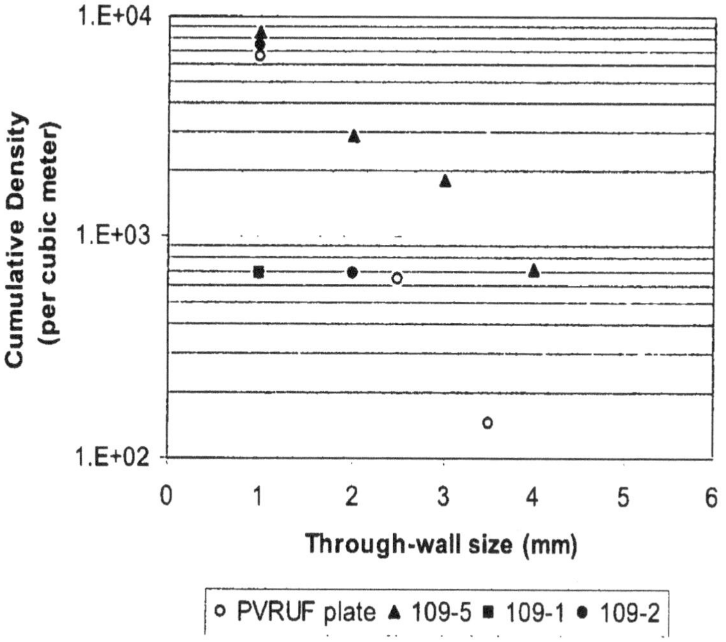

Figure 6 **Average of Validated Cumulative Flaw Density for Forging Material, A508**

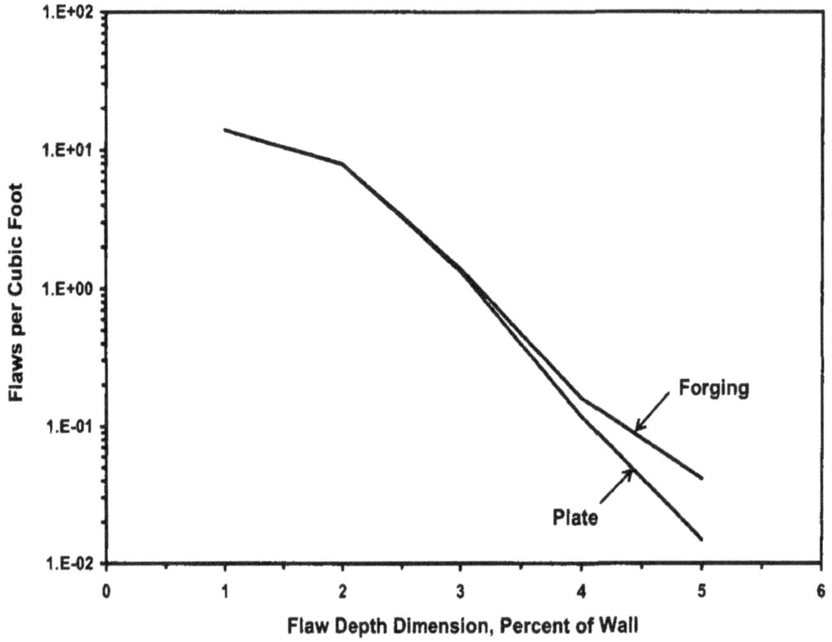

Figure 7 Comparison of Flaw Distributions for Forging and Plate

---

Basis for Assigning Subclad Flaw Distributions

F.A. Simonen
Pacific Northwest National Laboratory
Richland, Washington

September 29, 2004

Pacific Northwest National Laboratory (PNNL) has supported the U.S. Nuclear Regulatory Commission (USNRC) in the efforts to revise the Pressurized Thermal Shock (PTS) Rule. In this role PNNL has provided Oak Ridge National Laboratory (ORNL) with inputs to describe the distributions of fabrication flaws in reactor pressure vessels. These inputs, consisting of computer files, have been a key input to the probabilistic fracture mechanics code FAVOR. Flaw inputs have addressed seam welds, cladding and base metal materials, but had specifically excluded subclad flaws associated with the heat affected zone (HAZ) from the process that deposits stainless steel cladding to the inner surface of the vessel. Recently ORNL was requested by USNRS to evaluate the potential contribution of these subclad flaws to reactor pressure vessel failure for PTS conditions. The present paper describes the technical basis for the subclad flaw input files that PNNL provided to ORNL for use with the FAVOR code.

PNNL has examined material from vessels welds, basemetal and cladding and has used the data on observed flaws in these material types to establish statistical distributions for the numbers and sizes of flaws in these categories of materials. None of the examined material showed any evidence of the type of subclad flaws of interest. Therefore, the numbers and sizes of sub clad flaws for a vessel susceptible to such cracking was estimating from a review of the literature. The primary source was a comprehensive paper summarizing European work from the 1970's (A. Dhooge, R.E. Dolby, J. Sebille, R. Steinmetz ad A.G. Vinckier, "A Review of Work Related to Reheat Cracking in Nuclear Reactor Pressure Vessel Steels", International Journal of Pressure Vessels and Piping, Vol. 6, 1978, pp.329-409). This paper was based on experience with vessel cracking in Europe and subsequent research programs conducted during the 1970's. The paper should therefore be relevant to US concerns with older vessels that may have been fabricated with European practices.

The literature shows that subclad cracks 1) are shallow flaws extending into the vessel wall from the clad-to-basemetal interface with 4-mm being cited as a bounding through-wall depth dimension, 2) have orientations normal to the direction of welding for clad deposition - giving axial cracks in a vessel beltline, 3) occur as dense arrays of small cracks extending into the vessel wall from the clad to basemetal interface, 4) extend to depths limited by the heat affected zone. Pictures in the cited paper show networks of cracks with typical depths estimated from micrograph being significantly less than the bounding 4-mm depth. The cracks occur perpendicular to the direction of welding and are clustered where the passes of strip clad contact each other. Subclad flaws are much more likely to occur in particular grades of pressure vessel steels that have chemical compositions that enhance the likelihood of cracking. Forging grades such as A508 are more susceptible than plate materials such as A533. High levels of heat input during the cladding process also enhance the likelihood of subclad cracking. In addition other details of the cladding process are important such as single layer versus two layer cladding.

The numbers of cracks per unit area of vessel inner surface were estimated from Figure 1 of the Dhooge paper. Cracking was shown to occur in bands estimated to have a width of 4 mm. This dimension was used to estimate a bounding length of subclad cracks. The longest individual cracks in Figure 1 were about 2-mm versus the 4-mm width dimension of the zone of cracking. By counting the number of cracks pictured in small region of vessel surface crack density of 80,512 flaws per square meter was estimated.

The flaw input files as provide to ORNL were based on the following assumptions:

(1) The crack depth dimensions were described by a uniform statistical distribution from 0 to 2 mm with no cracks greater than 2 mm in depth.

(2) The crack lengths were also described by a uniform statistical distribution. Like our assumption for flaws in seam welds, the amount by which flaw lengths exceed their corresponding depth dimension is taken to be a uniform distribution from 0 to 4 mm.

(3) The flaw density expressed as flaws per unit area was converted (for purposes of the FAVOR code) to flaws per unit volume based on the total volume of the metal in the vessel wall.

(4) The file prepared for FAVOR assumes that the code simulates flaws for the total vessel wall thickness, rather than just the category 1 and 2 regions which address only the inner 3/8 of the wall thickness. Terry will need to account for this concern during the FAVOR calculations

The resulting very large number of flaws (> 130,000) per vessel is based on a photograph of one small area of a vessel surface, with the implication that this area was representative of the entire vessel. It is possible that subclad flaws tend to occur in patches of the vessel surface. However it is generally understood that subclad flaws occur in a wide spread manner and that there are very large numbers of flaws given the conditions for subclad cracking exist. Based on PNNL's limited review of documents it is difficult to justify reducing the estimated flaw density of subclad flaws. However, it would be useful to perform a sensitivity calculation to see if refinement of my estimate would have a significant effect on the FAVOR calculations.

NRC FORM 335
(9-2004)
NRCMD 3.7

U.S. NUCLEAR REGULATORY COMMISSION

BIBLIOGRAPHIC DATA SHEET

(See instructions on the reverse)

1. REPORT NUMBER
(Assigned by NRC, Add Vol., Supp., Rev., and Addendum Numbers, if any.)

NUREG-1806, Vol.2

2. TITLE AND SUBTITLE

Technical Basis for Revision of the Pressurized Thermal Shock (PTS) Screening Limit in the PTS Rule (10 CFR 50.61): Appendices

3. DATE REPORT PUBLISHED

MONTH	YEAR
August	2007

4. FIN OR GRANT NUMBER

5. AUTHOR(S)

Mark EricksonKirk, Mike Junge, William Arcieri, B. Richard Bass, Robert Beaton, David Bessette, T.H.(James) Chang, Terry Dickson, C. Don Fletcher, Alan Kolaczkowski, Shah Malik, Todd Mintz, Claud Pugh, Fredric Simonen, Nathan Siu, Donnie Whitehead, Paul Williams, Roy Woods, Sean Yin

6. TYPE OF REPORT

Technical

7. PERIOD COVERED *(Inclusive Dates)*

6-99 to 6-06

8. PERFORMING ORGANIZATION - NAME AND ADDRESS *(If NRC, provide Division, Office or Region, U.S. Nuclear Regulatory Commission, and mailing address; if contractor, provide name and mailing address.)*

Division of Fuel, Engineering, and Radiological Research
Office of Nuclear Regulatory Research
U.S. Nuclear Regulatory Commission
Washington, DC 20555-0001

9. SPONSORING ORGANIZATION - NAME AND ADDRESS *(If NRC, type "Same as above"; if contractor, provide NRC Division, Office or Region, U.S. Nuclear Regulatory Commission, and mailing address.)*

Division of Fuel, Engineering, and Radiological Research
Office of Nuclear Regulatory Research
U.S. Nuclear Regulatory Commission
Washington, DC 20555-0001

10. SUPPLEMENTARY NOTES

M. EricksonKirk, NRC Project Manager

11. ABSTRACT *(200 words or less)*

During plant operation, the walls of reactor pressure vessels (RPVs) are exposed to neutron radiation, resulting in localized embrittlement of the vessel steel and weld materials in the core area. If an embrittled RPV had a flaw of critical size and certain severe system transients were to occur, the flaw could very rapidly propagate through the vessel, resulting in a through-wall crack and challenging the integrity of the RPV. The severe transients of concem, known as pressurized thermal shock (PTS), are characterized by a rapid cooling of the internal RPV surface in combination with repressurization of the RPV. Advancements in our understanding and knowledge of materials behavior, our ability to realistically model plant systems and operational characteristics, and our ability to better evaluate PTS transients to estimate loads on vessel walls led the NRC to realize that the earlier analysis, conducted in the course of developing the PTS Rule in the 1980s, contained significant conservatisms.

This report summarizes 21 supporting documents that describe the procedures used and results obtained in the probabilistic risk assessment, thermal hydraulic, and probabilistic fracture mechanics studies conducted in support of this investigation. Recommendations on toughness-based screening criteria for PTS are provided.

12. KEY WORDS/DESCRIPTORS *(List words or phrases that will assist researchers in locating the report.)*

pressurized thermal shock, probabilistic fracture mechanics, nuclear reactor pressure vessel
pressurized water reactor

13. AVAILABILITY STATEMENT

unlimited

14. SECURITY CLASSIFICATION

(This Page)

unclassified

(This Report)

unclassified

15. NUMBER OF PAGES

16. PRICE

www.ingramcontent.com/pod-product-compliance
Lightning Source LLC
Chambersburg PA
CBHW081435170526
45166CB00008B/2206